ROBOT PROGRAMMING BY DEMONSTRATION: A PROBABILISTIC APPROACH

Engineering Sciences   *Micro- and Nanotechnology*

ROBOT PROGRAMMING BY DEMONSTRATION: A PROBABILISTIC APPROACH

Sylvain Calinon

EPFL Press
A Swiss academic publisher distributed by CRC Press

CRC Press
Taylor & Francis Group

Taylor and Francis Group, LLC
6000 Broken Sound Parkway, NW, Suite 300,
Boca Raton, FL 33487

Distribution and Customer Service
orders@crcpress.com

www.crcpress.com

Library of Congress Cataloging-in-Publication Data
A catalog record for this book is available from the Library of Congress.

The author and publisher express their thanks to the Swiss Federal Institute
of Technology in Lausanne (EPFL) for the generous support towards the
publication of this book.

This book is published under the editorial direction of
Professor Juergen Brugger (EPFL).

For the work described in this book, Dr. Silvain Calinon was awarded the "EPFL
Press Distinction", an official prize discerned annually at the Ecole polytechnique
fédérale de Lausanne (EPFL) and sponsored by the Presses polytechniques et
universitaires romandes. The Distinction is given to the author of a doctoral thesis
deemed to have outstanding editorial, instructive and scientific qualities; the
award consists of the publication of this present book.

is an imprint owned by Presses polytechniques et universitaires romandes, a Swiss
academic publishing company whose main purpose is to publish the teaching and
research works of the Ecole polytechnique fédérale de Lausanne (EPFL).

Presses polytechniques et universitaires romandes
EPFL – Centre Midi
Post office box 119
CH-1015 Lausanne, Switzerland
E-Mail: ppur@epfl.ch
Phone: 021 / 693 21 30
Fax: 021 / 693 40 27

www.epflpress.org

© 2009, First edition, EPFL Press
ISBN 978-2-940222-31-5 (EPFL Press)
ISBN 978-1-4398-0867-2 (CRC Press)

Printed in Italy

ACKNOWLEDGMENT

I express my gratitude to Professor Aude Billard for introducing me to the world of Robot Programming by Demonstration and for her continuous support and encouragement during my stay at the *Learning Algorithms and Systems Laboratory* (LASA) as a Ph.D. student and postdoctoral fellow, as well as to all my colleagues and friends there. This project would not have been possible without their contribution and support (it is for this reason that I utilize the "we" form throughout the book). I acknowledge the three examiners of my doctoral thesis, Professor Hervé Bourlard, Dr. Yiannis Demiris and Professor Stefan Schaal for providing me with their comments and advices on an early version of this manuscript, which was the principal source of inspiration to write this book. Thanks go to *EPFL-Press* for giving me the opportunity to publish this work and to the *Ecole Polytechnique Fédérale de Lausanne* for the chance to work in a great scientific environment and enjoyable atmosphere. A special thanks goes to Basilio Noris who designed the image of the cover. Last but not least, thanks go out to my family and my girlfriend for encouraging me during the writing phase.

The financial support for the research was provided by the *Secrétariat d'Etat à l'Education et à la Recherche Suisse* (SER), under Contract FP6-002020, Integrated Project *Cogniron* (http://www.cogniron.org) of the European Commission Division FP6-IST Future and Emerging Technologies, by the *Robot@CWE* European project (http://www.robot-at-cwe.eu) under contract FP6-2005-IST-5, by the *Feelix Growing* European project (http://www.feelix-growing.org) under contract FP6 IST-045169, and by the Swiss National Science Foundation, through grant 620-066127 of the SNF Professorships program.

CONTENTS

CHAPTER 1

INTRODUCTION

Since you are holding this book in your hands, it is very likely that you are interested in robots, artificial intelligence or technology-related applications. And even if this is not the case, there are chances that you use a computer more than once a week to facilitate your work (and your life in general). There is even greater chance that you do this through dedicated softwares. You may write a report, send an e-mail, check your budget on a spreadsheet, or process images from your digital camera, and doing so, you rely on existing user-friendly applications. Even if you are a programmer or know the technical details of how a computer works, you will start most of the tasks you want to achieve with the computer by opening the proper application instead of reprogramming the machine from scratch. This type of computer use seems obvious today, but at the beginning of the computer age, it was considered as obligatory and evident to be familiar with a programming language in order to make use of such a machine. From the apparition of the first personal computers, much time went by before end-users could finally do without programming skills to interact efficiently with their computers through simple (or at least quite simple) interfaces.

Performing tasks with a computer without knowing how its processor works sounds obvious today, but in the field of robotics, this path is unfortunately followed very slowly. A large amount of people today will not negate that robots are very useful in a wide range of applications, but many of them still do not conceive that these robots need to be used and reprogrammed by end-users as easily as computers. In other words, the common thinking is that, to use a robot to facilitate your task, you need to know how to program it.

Creating user-friendly systems to teach robots new skills is a challenging task that is at core of numerous research activities aiming at demonstrating that robots will probably have their place in our homes and offices in a couple of years.[1] To accomplish this, we may take inspiration from the development of user-friendly applications in computer science, but we cannot simply

---

[1] At the end of 2007, about 1 million industrial robots and 5.5 million service robots were, worldwide, operating in factories, in dangerous or tedious environments, in hospitals, in private homes, in public buildings, underwater, underground, on fields, in the air or in

re-apply this knowledge to robotics. Indeed, compared to a computer, the robot moves in its environment. Taking this embodiment into account is a key issue when designing the controllers required to interact with these robots. Compared to the use of a fixed 2-dimensional computer screen, creating applications for robots involves simultaneously controlling several degrees of freedom with versatile methods in order for the robot to be used in very different environments.

Robot Programming by Demonstration (PbD) covers methods by which a robot learns new skills through human guidance. Also referred to as learning by imitation or apprenticeship learning, PbD takes inspiration from the way humans learn new skills by imitation, thereby developing methods by which new skills can be transmitted to a robot. PbD covers a broad range of applications. In industrial robotics, the goal is to reduce the time and costs required to program the robot. The rationale is that PbD would allow the modification of an existing product, the creation of several versions of a similar product or the assembly of new products in a very rapid way, and this could be done by lay users without help from an expert in robotics. PbD is perceived as particularly useful when it comes to service robots, i.e., robots deemed to work in direct collaboration with humans. In this case, methods for PbD go beyond transferring skills and offer new ways for the robot to interact with the human, from being capable of recognizing people's motion to predicting their intention and seconding them in the accomplishment of complex tasks. As the technology has improved to provide these robots with more and more complex hardware, including multiple sensor modalities and numerous degrees of freedom, robot control and especially robot learning have become increasingly complex. Learning control strategies for numerous degrees of freedom platforms deemed to interact in complex and variable environments, such as households, is faced with two key challenges: first, the complexity of the tasks to be learned is such that pure trial-and-error learning would be too slow. PbD thus appears to be a good approach for speeding up learning by reducing the search space, while still allowing the robot to refine its model of the demonstration through trial and error. Second, there should be a continuum between learning and control, so that control strategies can adapt on the fly to drastic changes in the environment. The present work addresses both challenges in investigating methods by which PbD is used to learn the dynamics of robot motion, and, by so doing, provide the robot with a generic and adaptive model of control.

Robot PbD is now a regular topic at the two major conferences on robotics (IROS and ICRA), as well as at conferences on related fields, such as Human-Robot Interaction (HRI, AISB) and Biomimetic Robotics (BIOROB, HUMANOIDS).

---

space. In countries such as Japan, nearly 3% of the manufacturing workers were at this time robots (International Federation of Robotics (IFR), 2008).

1.1   CONTRIBUTIONS

This book focuses on the generic questions addressed by the *what-to-imitate* and *how-to-imitate* issues, which are respectively concerned with the problem of extracting what are the essential features of a task and determining which way to reproduce these essential features under varying conditions. The perspective adopted is that multiple demonstrations of a similar task can help the robot *understand* the key aspects of the skill. In other words, when a skill shares similar components across multiple demonstrations, it might suggest that these components are essential features that should be reproduced by the robot. After considering this, we suggest the use of various probabilistic methods for analyzing a task at a trajectory level. The statistical models presented in this book have already been employed in various fields of research but their use in PbD is new. It requires the adaption of the algorithms to the constraints inherent to the specificities of the paradigm and the practical testing of whether these models can be used and combined efficiently in *Learning by Imitation* scenarios. Moreover, these probabilistic models have essentially been described for recognition purposes, leaving out the reproduction capabilities of the models. The use of these models to reproduce new skills by imitation is thus challenging and has not been extensively explored.

The contributions of this book are threefold: (1) it contributes to PbD by proposing a generic probabilistic framework to deal with recognition, generalization, reproduction and evaluation issues, and more specifically to deal with the automatic extraction of task constraints and with the determination of a controller that is robust to perturbations and able to generalize a skill to varying contexts; (2) it contributes to *Human-Robot Interaction* (HRI) by proposing active teaching methods that place the human teacher "in the loop" of the robot's learning by using an incremental scaffolding process and different modalities to perform the demonstrations; (3) it finally contributes to robotics and HRI through various real-world applications of the framework, displaying learning and reproduction of communicative gestures and manipulation skills.

Source codes of the algorithms presented in this book are available online, and the experiments are exclusively conducted using robots and sensory hardware from companies that sell their products. Thus, the applications presented throughout the book are reproducible by other researchers and new applications can be built upon them. The book also demonstrates the generality of the learning framework by demonstrating that the hardware components can be easily changed when a new technology (or a new version of a product) is available, without having to apply major alterations to the framework.

1.2   ORGANIZATION OF THE BOOK

This book is decomposed into five main parts. The first chapter presents a historical overview of robot programming by demonstration and a recital of current state of the art. Chapter 2 presents different probabilistic models from a theoretical perspective and discusses their use within a PbD framework. Chapter 3 presents preliminary experiments for testing and optimizing the parameters of the proposed methods to stimulate their use in further PbD experiments. Chapters 4-7 gather the core of the experiments where the proposed probabilistic framework is applied to various human-robot teaching interactions. Various PbD issues are didactically illustrated through these experiments by gradually increasing the complexity of the system. Finally, Chapter 8 discusses the capabilities and limitations of the system and presents further directions of research.

1.3 REVIEW OF ROBOT PROGRAMMING BY DEMONSTRATION (PBD)

1.3.1 The birth of programmable machines

Providing the possibility to program and re-program a machine very quickly became the most important requirement for efficiently utilizing this machine. The idea of programming machines appeared with mechanical automata long before the apparition of computers or modern robots,[2] and it is difficult to determine when the first programmable automaton came into existence.

Heron of Alexandria, AD 10–70, (M. E. Rosheim, 1994) is one of the greatest experimenters of antiquity, having invented numerous user-configurable automated devices. He may be more famous as a mathematician than as an engineer, but some of his inventions are extraordinarily visionary in robotics.[3] His *Automaton Theater* indeed describes theatrical constructions that moved by means of strings wrapped around rotating drums animating figures that acted out a series of dramatic events. Energy was provided by a weight resting on a hopper full of grain, which leaked out through a small hole in the bottom. As the weight gradually sank, it pulled a rope wound around an axle of the stand to turn its wheels, thereby creating movement. Along with the power of falling weights, these figures used the basic mechanical resources of wheels, pulleys, and levers to create a variety of motion by delivering power over a relatively long period.

Despite the possibility of fine-tuning the device by hand, it remains difficult to determine whether the mechanism was aimed at being reprogrammable to perform other plays. The first automaton that was reprogrammed every day can probably be credited to the Muslim inventor Al-Jazari who fabricated one of the earliest forms of programmable humanoid robots. His *Castle Water Clock*, described in 1206, was an elaborate machine designed to keep time

---

[2] The word robot was popularized by the Czech author Karel Capek in his 1921 play R.U.R. (Rossum's Universal Robots).

[3] He may also be behind the invention of the first feedback control device with his self-filling wine bowl, which had a hidden float valve that automatically sensed the level of wine in a bowl and that mysteriously refilled itself (Mayr, 1970).

(Hill, 1973). At regular intervals throughout the day, the machine was designed to carry out specific actions such as opening doors, letting falcons drop balls and actuating automaton musicians to show the passing hours. Behind the clock was an ingenious mechanism driving this sophisticated chain of events. Water drained out of a vertical tube under gravity to provide power to turn cams on a shaft, thus actuating the events. The medieval Islamic day began at sunset and the hours were counted between sunset and the end of twilight. Since the days became longer in summer or shorter in winter, the automaton had to be re-programmed to match the different lengths of hours, which was done every day.

The first self-propelled programmable robot can probably be attributed to Leonardo da Vinci during the Renaissance, with a mechanical lion robot offered in 1515 to Francis I, King of France. The automaton was said to be capable of walking a few steps across the floor, rising on its hindquarters, opening its chest and then delivering flowers (M. Rosheim, 2006). In 1495, Leonardo also designed a humanoid automaton in knight's armor to entertain. This automaton was supposedly able to sit up, wave its arms and move its head, but it remains unknown if the design was ever built.

The first truly modern, digitally operated and programmable robot was invented by George Devol and Joseph Engelberger in 1954 and was called Unimate (Waurzyniak, 2006). This 4000-pound robotic arm was bought by General Motors and installed in 1961 in a plant in New Jersey, obeying step-by-step commands stored on a magnetic drum, in order to lift hot pieces of metal from a die casting machine and stacking them.

1.3.2 Early work of PbD in software development

Programming by Example (PbE) or *Programming by Demonstration* (PbD) appeared both in robotics and software development research in the early 80's. In software development, PbD provided an intuitive and flexible way for the end-user to re-program a computer simulation without having to learn a computer language (Cypher et al., 1993; Smith, Cypher, & Spohrer, 1994; Mitchell, Caruana, Freitag, McDermott, & Zabowski, 1994). The idea emerged by observing that most of the end-users were skilled at using a computer and its associated interfaces and applications (edition skills) but were only able to re-program them in a limited way (by setting preferences). Due to the restricted impact of the numerous attempts at creating simple languages for the end-users to re-program their computers (e.g., the LOGO programming language), certain researchers focused on re-formulating the problem from its source. Instead of designing an improved syntax for a programming language, they took the perspective that the computer could learn differently by taking into consideration that most computer applications follow similar graphical user interface strategies (environments with windows, menus, icons and a pointer controlled by the mouse). By taking advantages of the user's knowledge of the task (i.e., the fact that the user knows how to perform a task but not how to program it), they designed softwares that were able to extract rules simply by using the interfaces (Cypher et al., 1993; Lieberman, 2001). The syntax of the program was then hidden from the users who were only demonstrating the skill to the computer. Due to the restricted behaviors that can be applied to the considered graphical environment, the learned skills consisted mostly of discrete sets of "if-then" rules that were automatically extracted by demonstrating the tasks to the computer. It was thus possible to generalize the skill and reproduce the behavior under other circumstances by sharing similarities with the learned skill (Smith et al., 1994). For example, Mitchell et al. (1994) proposed to use *feed-forward neural networks* to learn the user's preferences when utilizing a calendar, where the application was progressively able to automatically plan adequate meetings and appointments with respect to the end-user's needs.

1.3.3   Early work of PbD in robotics

At the beginning of the 1980s, PbD started attracting attention in the field of manufacturing robotics. It appeared as a promising route for automating the tedious manual programming of robots and as a way of reducing the costs involved in the development and maintenance of robots in a factory.

An operator has implicit knowledge of the task to achieve. He/she knows how to do it but does not have the necessary programming skills or the time required to reconfigure the robot. Demonstrating how to achieve the task through examples would thus permit learning the skill without explicitly programming each detail. The PbD paradigm in robotics, however, generated new perspectives as opposed to PbD in software development by allowing the demonstrations and the reproduction attempts to be performed on different media. Indeed, when considering PbD in software development, the demonstrator shares the same *embodiment* as the one of the imitator (the application on the 2D computer screen). In contrast, by using different architectures to demonstrate and reproduce the skills (as may be the case for robotics applications), the issues related to the difference in *embodiments* are also introduced, later referred to as *correspondence problems*.

As a first approach to PbD, symbolic reasoning was commonly adopted in robotics with processes referred to as *teach-in*, *guiding* or *play-back* methods (Lozano-Perez, 1983; Dufay & Latombe, 1984; Levas & Selfridge, 1984; Segre & DeJong, 1985; Segre, 1988). In these studies, PbD was performed through manual (teleoperated) control. The position of the end-effector and the forces applied on the manipulated object were stored throughout the demonstrations together with the positions and orientations of the obstacles and of the target. This sensorimotor information was then segmented into discrete subgoals (keypoints along a trajectory, see Fig. 1.1) and into appropriate primitive actions to attain these subgoals. Primitive actions were commonly chosen to be simple point-to-point movements employed by the industrial robots at this time. Examples of subgoals included for instance the robot's gripper orientation and position in relation to the goal (Levas & Selfridge, 1984). Consequently, the demonstrated task was segmented into a sequence of state-action-state transitions.

To take into account the variability of human motion and the noise inherent to the sensors capturing the movements, it appeared necessary to develop

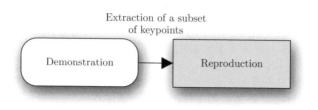

Figure 1.1 The exact copy of a skill by interpolating between a set of pre-defined keypoints, see for instance (Lozano-Perez, 1983).

a method that would consolidate all demonstrated movements. For this purpose, the state-action-state sequence was converted into symbolic "if-then" rules, describing the states and the actions according to symbolic relationships, such as "in contact", "close-to", "move-to", "grasp-object", "move-above", etc. Appropriate numerical definitions of these symbols (i.e., at what distance an object would be considered as "close-to" or "far-from") were given as prior knowledge to the system. A complete demonstration was thus encoded in a graph-based representation, where each state constituted a graph node and each action a directed link between two nodes. Symbolic reasoning could subsequently unify different graphical representations for the same task by merging and deleting nodes (Dufay & Latombe, 1984).

1.3.4 Toward the use of machine learning techniques in PbD

The PbD methods presented above still employed direct repetition, which was useful in automation to reproduce an exact copy of the motion. In order to apply this concept to products with different variants or to transfer the programs to new robots, the generalization issue became crucial. To address it, the first attempts at generalizing the skill were mainly based on the help of the user through queries about his intentions (Heise, 1989; Friedrich, Muench, Dillmann, Bocionek, & Sassin, 1996). Following this, various levels of abstractions were proposed to resolve the generalization issue, basically dichotomized in learning methods at a *symbolic level* or at a *trajectory level*. A task at a *symbolic level* is described by the sequential or hierarchical organization of a discrete set of primitives that are pre-determined or extracted with pre-defined rules. A task at a *trajectory level*, on the other hand, is described by temporally continuous signals representing different configuration properties changing over time (e.g., current, torque, position, orientation). Different levels of abstraction can be utilized (the position of the end-effector, for instance, is a representation of a higher level than the joint angle trajectories, which are also of higher level as opposed to directly considering the current sent to the motors to set the robot to this specific posture).

Machine Learning (ML) appeared to be an appealing solution for dealing with the generalization issue (Thrun & Mitchell, 1993; Heise, 1989; Friedrich et al., 1996). Several techniques proposed in ML could be tested with multiple examples of input/output (sensory/actuators) datasets that fit most of the frameworks developed in ML perfectly, while robotics benefitted from the ML ability to cope with multivariate data and generalization capabilities. It was thus possible to extend the *record/replay* process to one of generalization.

In industrial robotics, Muench, Kreuziger, Kaiser, and Dillmann (1994) suggested the use of ML algorithms to analyze *Elementary Operators* (EOs), defining a discrete set of basic motor skills. Already in this early work, the authors pointed to several key problems of ML in robotics – which are still not completely solved today – in terms of encoding, generalization, reproduction of a skill in new situations, evaluation of a reproduction attempt, and role of the user in the learning paradigm. First they introduced the problems related to the segmentation and extraction of EOs, which was resolved in their framework through the user's support. To leverage this task, they suggested the use of statistical analysis to maintain only the EOs that appeared repeatedly (generalization at a symbolic level). Then, by observing several sequences of EOs, they pointed to the relevance of extracting a structure from these sequences (generalization at a structured level). In this early work, both generalization processes were supervised by the user, who is asked to account for whether an observation is example- or task-specific. Muench et al. (1994) admitted that generalizing over a sequence of discrete actions was only one part of the problem since the controller of the robot also required the learning of continuous functions in order to control the actuators. They proposed to overcome the missing parts of the learning process by leveraging them to the user, who took an active role in it. They highlighted the importance of providing a set of examples that are usable by the robot: (1) by constraining the demonstrations to modalities that the robot can understand; and (2) by providing a sufficient number of examples to achieve a desired generality. They noted the importance of providing an adaptive controller to reproduce the task in new situations, that is, how to adjust an already acquired program. The evaluation of a reproduction attempt was also leveraged to the user by letting him/her provide additional examples of the skill in the regions of the learning space that had not yet been covered. In this way, the teacher/expert could control the generalization capabilities of the robot.[4] Even if research in PbD was turned toward a fully autonomous learning system, early results quickly showed the importance and advantages of using an interaction process involving the user and the robot to cope with the deficiencies of the current learning systems. This extraction of the dependencies at a symbolic level was also referred to as a *functional induction* problem in Dufay and Latombe (1984), still explored by many researchers today (Saunders, Nehaniv, & Dautenhahn, 2006; Ekvall & Kragic, 2006; Pardowitz, Zoellner, Knoop, & Dillmann, 2007; Steil, Röthling, Haschke, & Ritter, 2004; Nicolescu & Mataric, 2003).

---

[4] Note that the term "expert" can be misleading since it makes no distinction between an *expert at executing* the skill and an *expert at teaching*. The first definition involves that the user has skills for the execution of the task considered while the second involves pedagogical skills.

To summarize, early work in PbD principally adopted a user-guided generalization strategy by querying the user to provide additional sources of information for the induction process. The approach presented in this book consists in leveraging this process (or at least modifying the user's role to avoid a burden of queries from the robot) by applying and combining several probabilistic ML algorithms initially developed for large datasets to a PbD framework where the training dataset is limited to the few demonstrations provided to the robot.

1.3.5 From a simple copy to the generalization of a skill

The field progressively moved from simply copying the demonstrated movements to generalizing across sets of demonstrations.

The analogy with *motor tapes*, initially proposed by Atkeson et al. (2000), consisting in the robot storing an explicit representation of a movement trajectory in memory and executing the appropriate tape to reproduce a task, was progressively extended to more sophisticated versions where a combination of a set of tapes was used to produce the final movement. To depart from a user-guided generalization strategy, in which the robot may ask the user extensively for additional sources of information, PbD started incorporating more of ML tools to tackle both the perception issue, i.e., how to generalize across demonstrations, and the production issue, i.e., how to generalize the movement to new situations. These tools include for example *Artificial Neural Networks* (ANNs) (Liu & Asada, 1993; Billard & Hayes, 1999), *Radial-Basis Function Networks* (RBFs) (Kaiser & Dillmann, 1996), *Fuzzy Logic* (Dillmann, Kaiser, & Ude, 1995), and *Hidden Markov Models* (HMMs) (Yang, Xu, & Chen, 1993; Pook & Ballard, 1993; Hovland, Sikka, & McCarragher, 1996; Tso & Liu, 1996; C. Lee & Xu, 1996).

The idea was born that learning through self-experimentation and practice could be used jointly with imitation, where observations are used to initiate further search of a controller by repeated reproduction attempts. In Atkeson and Schaal (1997), a single observation initiated the further search of a controller solution by repeated reproduction attempts. The authors demonstrated their approach with a robot manipulator, learning how to swing a pendulum by first observing the user performing the task and then updating the model through practice. In Bentivegna, Atkeson, and Cheng (2004), a humanoid robot learned how to play air-hockey and how to move a ball in a pan-tilt maze game. The sub-goals of a skill were first extracted by observation and the robot learned how to achieve these sub-goals by practicing the task through reinforcement learning strategies (Peters, Vijayakumar, & Schaal, 2003; Yoshikai, Otake, Mizuuchi, Inaba, & Inoue, 2004). A body of work was also dedicated to the individual exploration of the search space as a basis for investigating how knowledge is propagated among a population of agents (Jansen & Belpaeme, 2006; Belpaeme, Boer, & Jansen, 2007; Steels, 2001).

Another trend of research, which is the one adopted in this book, took the perspective that observing multiple demonstrations can help generalize a skill by extracting the requisites of the task. In order to extract and reproduce only the essential aspects of a task, various approaches based on statistics were proposed. Thus invariant features across multiple demonstrations could be extracted. For instance, Ude (1993) used spline smoothing techniques to deal with the uncertainty contained in several demonstrations performed in a *joint space* or in a *task space*. In Delson and West (1996), the amount of human variation among multiple trials was used to define the accuracy requirements along a motion in a *Pick & Place* task. The different trajectories formed a boundary region defining a range of acceptable trajectories where low variation indicates requirements of high accuracy. In Ogawara, Takamatsu, Kimura, and Ikeuchi (2003), the robot acquired a set of sensory variables while demonstrating a manipulation task consisting in arranging various objects. At each time step, the robot stored and computed the mean and variance of the collected variables. The sequence of means and associated variances could then be used as a simple generalization process, respectively providing a generalized trajectory and associated constraints.

1.3.6   From industrial robots to service robots and humanoids

Building a human-like artificial creature is a longstanding scientist dream – see for example the humanoid automaton designed by Leonardo da Vinci during the 15th century (M. Rosheim, 2006) – which has been naturally extended by robotics engineers since the *Waseda University* introduced *WABOT-1* in 1973 as the first full-scaled humanoid robot (Kato, 1973). Beside the intriguing nature and technical challenge of humanoids, their introduction in robotics research have led to challenges in terms of learning and communication, where the robot is expected to collaborate with humans in a shared environment. Compared to industrial robots that are often programmed to carry out tasks in static environments, humanoid robots are expected to carry out large panels of skills in dynamic situations. Instead of re-designing the environment for the robot, the robot architecture and learning components are re-designed to adapt to changing environments, i.e., the roles of the hardware (robot) and software (learning) components are conceptually recast.

Besides the ergonomic benefit of a humanoid shape moving in human surroundings, the architecture of these robots is also optimized for communication, which increases its capabilities for teamwork and companionship.[5] New challenges for learning in robotics came along with these new architectures. Since the humanoids need to adapt to new environments, the algorithms used for their control require flexibility and versatility. Due to the continuously changing environments and to the huge varieties of tasks that the robot is expected to perform, it requires the ability to continuously learn new skills and adapt existing skills to new contexts. Indeed, the robot is expected to physically cooperate with the human user (e.g., by moving in the same rooms, sharing the same devices or manipulating the same tools), and must thus be predictable

---

[5] Note that the social dimension of the interaction can take on a variety of cultural forms (Kaplan, 2004).

and behave in a human-like manner with regard to social interaction, gestures or learning behavior.

1.3.7 From a purely engineering perspective to an interdisciplinary approach

With an increasing development of mobile and humanoid robots that are more animal-like in their behaviors, the field has become directed towards adopting an interdisciplinary approach. PbD research departs progressively from its purely engineering perspective, gaining insights from neuroscience and social sciences, to emulate the process of imitation in humans and animals. The first insight was particularly marked with the discovering in primates of specific neural mechanisms for visuo-motor imitation (referred to as mirror neurons) (Rizzolatti, Fogassi, & Gallese, 2001; Rizzolatti, Fadiga, Fogassi, & Gallese, 2002; Decety, Chaminade, Grezes, & Meltzoff, 2002). These studies have inspired a large trend of research aiming at finding biologically plausible solutions to the PbD paradigm (Billard & Schaal, 2006; Arbib, Billard, Iacobonic, & Oztop, 2000; Y. Demiris & Hayes, 2002). The second insight was marked by a collection of work exploring the developmental stages of imitation capabilities in children (Piaget, 1962; Meltzoff & Moore, 1977; Bekkering, Wohlschlaeger, & Gattis, 2000; Nadel, Guerini, Peze, & Rivet, 1999; Zukow-Goldring, 2004). With the increasing consideration of this body of work in robotics, the notion of *Robot Programming by Demonstration* was eventually replaced by the more biological labeling of *Learning by Imitation* (Billard & Dillmann, 2006).

Novel learning challenges could, thus, be set forth. Robots were expected to show a high degree of flexibility and versatility, both in their learning system and in their control system, in order to be able to naturally interact with human users and demonstrate similar skills (e.g., by employing identical infrastructures and manipulating the same tools as humans). Robots were more and more expected to act "human-like" in order for their behavior to be more predictable and acceptable. Thanks to this swift bioinspiration, Robot PbD became once again a core topic of research in robotics (Matarić, 2002; Schaal, 1999b; Billard, 2002; Dautenhahn, 1995), and this after the original wave of robotic imitation based on symbolic artificial intelligence methods lost its thrust in the late 1980s.

The field identified a number of key problems that needed to be Solved in order to ensure such a generic approach for transferring skills across various agents and situations (Nehaniv & Dautenhahn, 2000; Nehaniv, 2003). These were formulated as a set of generic questions, namely *what to imitate*, *how to imitate*, *when to imitate* and *who to imitate*, which were expressed in response to the large body of diverse work in Robot PbD, not easily unified under a small number of coherent operating principles (Bakker & Kuniyoshi, 1996; Ehrenmann, Rogalla, Zoellner, & Dillmann, 2001; Kaiser & Dillmann, 1996; Skubic & Volz, 2000; Yeasin & Chaudhuri, 2000; J. Zhang & Rössler, 2004). The above four questions and their solutions aim at being generic in the sense that no assumptions are made regarding the type of skills that may be transmitted.

1.4   CURRENT STATE OF THE ART IN PBD

The previous section described the various approaches adopted in PbD by providing a principally historical overview. In this section, we develop some of the aspects previously discussed and present the manner in which the approach proposed in this book has a place in current research. We review the major trends of research by focusing essentially on work with direct applications to robotics.

1.4.1   Human-robot interfaces

One research trend focused on the sensory system of the robot by assuming that a robot should first be able to perceive and interpret a wide range of communicative modalities and cues in order to learn new skills. Among the various human-robot interfaces, those based on gesture are of particular interest to ensure the transfer of manipulation skill to the robot.

Mouses and keyboards are input devices that are very common in *Human Computer Interaction* (HCI) to control a software on a screen interface, i.e., to control a motion of 2 degrees of freedom (DOFs) or a sequence of symbolic commands. However, these input devices are not well designed for controlling the high number of DOFs of a robot, particularly when considering humanoids. For the transfer of skills to these robots, research in human-robot interfaces have focused on the simultaneous control of a high number of DOFs, for instance by using solutions originally developed for virtual reality or 3D computer graphics applications.

Interfaces for body gesture tracking can be roughly dichotomized into vision-based and non-vision-based systems. The former are often assumed to provide the most natural means of communicating between human and machine collaborators. However, this type of interaction depends considerably on the progress in the computer vision field (Shon, Storz, & Rao, 2007; D. Lee & Nakamura, 2007).

Information concerning human body gesture can also be retrieved by various mechanical, inertia or magnetic sensors attached to different body parts, recording position and/or orientation information that allows to reconstruct the evolution of the body pose through time. These sensors use and combine a variety of modalities to record information, commonly based on gyroscopes, accelerometers or magnetometers (see Hightower and Borriello (2001) and Rolland, Davis, and Baillot (2001) for surveys on tracking technologies). Compared to vision based trackers, motion sensors have the advantage of being robust to occlusion and to provide position and orientation information at a usually lower computational cost as opposed to vision. The main drawback of these non-vision-based devices is that the sensors must be attached to the human body and usually require a calibration phase. One should note that similar requirements also appear when using visual markers, and that the recent development of clothes directly encapsulating the sensors can lower this constraint (Rossi, Bartalesi, Lorussi, Tognetti, & Zupone, 2006).

Mechanically based trackers consist of a rigid articulated structure attached to the body of the user with encoders at the joint level (Hollerbach & Jacobsen, 1996; Tachi, Maeda, Hirata, & Hoshino, 1994). Similarly, an external robotic tracker (i.e., one not directly attached to the body of the user) can be employed to teach a robot new skills by demonstrating a gesture. Arm exoskeletons mainly utilized for re-education can also be used as a means of controlling robots by executing gestures (Gupta & O'Malley, 2006). The term *kinesthetic teaching* is used throughout this book to describe this process, achieved either by setting the robot's motors in a passive mode (Peshkin, Colgate, & Moore, 1996), or by implementing a compliant controller to move the robot, as if its arm was of very low weight and as if its articulation comprised no motors (Heinzmann & Zelinsky, 1999; Massie & Salisbury, 1994). One common solution is to use force sensors to perceive the force applied to the end-effector of the robot's arm. Haptic interfaces that convey force and torque feedback are widely used in telerobotics to dynamically interact with robotic agents operating on remote sites (Yokoi et al., 2006).

Human hands are capable of producing a wide range of poses due to their high number of articulations and they can deliver robust information for interaction (e.g., sign language, pointing). Similarly, hand gestures can convey significant information for human-robot interaction applications through data gloves designed to record hands postures, based principally on optical or resistive properties of flexible materials (bend-sensing devices). A survey of glove-based input interfaces can be found in (Sturman & Zeltzer, 1994). Applications in robotics range from the control of mobile robots by the recognition of specific hand gestures (Iba, Weghe, Paredis, & Khosla, 1999) via the use of data gloves in order to demonstrate a manipulation skill to a manipulator robot (Dillmann, 2004), to the learning of several dexterous types of grasping depending on the object to be manipulated (Caurin, Albuquerque, & Mirandola, 2004; Aleotti & Caselli, 2006a; Ekvall & Kragic, 2005).

Another way of providing information to the robot without relying on symbolic/verbal cues consists in carrying out part of the training in *Virtual Reality* (VR) or *Augmented Reality* (AR) and providing the robot with *virtual fixtures* (Kheddar, 2001; Dillmann, 2004; Aleotti, Caselli, & Reggiani, 2004; Ekvall & Kragic, 2005). This intermediate layer of interaction allows the user to concentrate on the task to achieve by leveraging the need of considering the specific capabilities of the robot (see Fig. 1.4). This interface can thus act as a flexible control space, providing information and feedback prior to the real execution on the robot, or in other words, by considering a bilateral teleoperation where the experience of the robot in the environment is reflected back to the operator (Yoon, Suehiro, Onda, & Kitagaki, 2006; Takubo, Inoue, Arai, & Nishii, 2006). An overview of the synergy between virtual reality and robotics has been presented elsewhere (Burdea, 1999; Nguyen et al., 2001). In Robot Programming by Demonstration, a solution for the transfer of a skill thus consists in performing the learning phase in the virtual world and the reproduction phase on the real robot (Dillmann, 2004; Aleotti et al., 2004). To accomplish this, the robot is usually modeled in a virtual space, where virtual fixtures (or virtual guides) are used to simplify and highlight the task requirements for the human operator (Kheddar, 2001; Otmane, Mallem, Kheddar, & Chavand, 2000). A direct extension consists in using *Augmented Reality* (AR) to display a real view of the robot space (e.g., by visualizing the robot on a video) where additional information is added to the image, thus allowing the user to perform a more efficient transfer of the skills (Milgram, Rastogi, & Grodski, 1995; Young, Xin, & Sharlin, 2007). A review of AR techniques can be found in Azuma et al. (2001). In Cannon and Thomas (1997), AR is used to collaboratively plan the execution of a robot action in a hazardous environment by allowing multiple users with varying fields of expertise to virtually visualize the effects of the motion before controlling the real robot. Bejczy, Kim, and Venema (1990) provided a graphical representation of the robot superimposed to the video of the real robot (called a phantom robot) for planning the further steps of the task.

1.4.2   Learning skills

As mentioned in Section 1.3, early approaches to solving the problem of how to generalize a given skill to a new/unseen context consists in explicitly asking the teacher for further information (Fig. 1.3). Additionally, further methods of

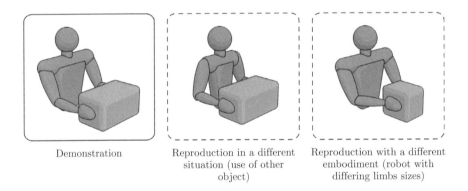

Demonstration   Reproduction in a different   Reproduction with a different
                     situation (use of other   embodiment (robot with
                     object)   differing limbs sizes)

Figure 1.2 Illustrations of the correspondence problems.

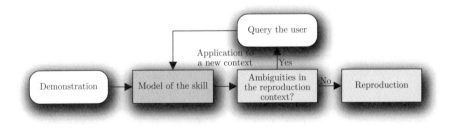

Figure 1.3 The learning of a skill through a query-based approach, see e.g., Muench et al. (1994).

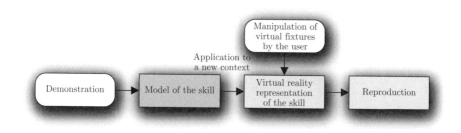

Figure 1.4 PbD in a Virtual Reality (VR) setup providing the robot with virtual fixtures. VR acts as an intermediate layer of interaction to complement real-world demonstration and reproduction.

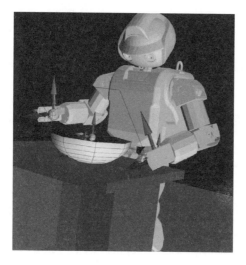

Figure 1.5 An example of the use of virtual fixtures to facilitate the teaching process in a Virtual Reality setup.

learning a skill include allowing a robot to automatically extract the important features characterizing the skill and to seek a controller that optimizes the reproduction of these characteristic features. Determining a *metric of imitation performance* is a key concept at the bottom of these approaches. One must first determine the metric, i.e., find the weights to be attach to reproducing each of the components of the skill. Once this is done, one can find an optimal controller to imitate by trying to minimize this metric (e.g., by evaluating several reproduction attempts or by deriving the metric to obtain an optimum). The metric acts as a cost function for the reproduction of the skill (Nehaniv & Dautenhahn, 2000). In other terms, a metric of imitation provides a means of quantitatively expressing the user's intentions during the demonstrations and of evaluating the robot's faithfulness at reproducing these. Figure 1.6 presents an illustration of the concept of a metric of imitation performance as well as its use to drive the robot's reproduction.

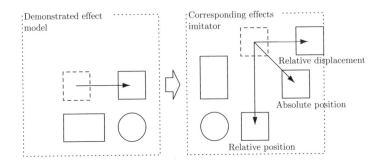

Figure 1.6 The use of a metric of imitation performance to evaluate a reproduction attempt and find an optimal controller for the reproduction of a task (here, to displace the square in a 2D world) (Alissandrakis et al., 2006).

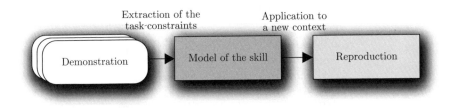

Figure 1.7 The generalization of a skill by extracting the statistical regularities across multiple observations, see e.g., Delson and West (1996).

In order to learn the metric (i.e., to infer the task constraints), one common approach consists in creating a model of the skill based on several demonstrations of the same skill performed in a slightly different context (Fig. 1.7). This generalization process resides in exploiting the variability inherent to the various demonstrations and extracting the essential components of the task. These should constitute the components that remain unchanged across the various demonstrations.

The problem of determining an optimal human motion representation has been addressed in various fields of research, including computer graphics (Brand & Hertzmann, 2000), biomechanics (Wu et al., 2005) and robotics (Gams & Lenarcic, 2006; Lenarcic & Stanisic, 2003). This issue plays a crucial role in understanding the mechanics of the human body and the dynamics of human motion. Engineering-oriented and Machine Learning approaches to robot PbD focus on developing algorithms that are generic in their representation of the skills as well as in the way they are generated. Current approaches for representing a skill can be broadly divided into two trends: a low-level representation of the skill, taking the form of a non-linear mapping between sensory and motor information, which we later refer to as "trajectories encoding", and, a high-level representation that decomposes the skill in a sequence of action-perception units, which we classify as "symbolic encoding". Figure 1.8 presents a schematic of the learning process in which a skill is represented at

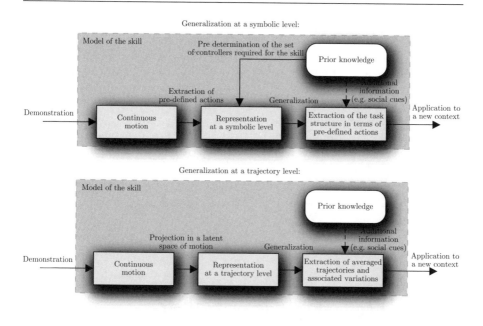

Figure 1.8 An illustration of the varying levels of representation when describing the skill.

Table 1.1 Advantages and drawbacks of representing a skill at a symbolic/trajectory level.

	Span of generalization	Advantages	Drawbacks
Symbolic level	Sequential organization of pre-defined motion elements	Allows to learn hierarchy, rules and loops	Requires the predefinition of a set of basic controllers for reproduction
Trajectory level	Generalization of movements	Generic representation of motion that allows encoding of very different types of signals/gestures	Does not allow the reproduction of complicated high-level skills

a symbolic or a trajectory level (these two schemas are detailed versions of the tinted boxes depicted in Fig. 1.7). Table 1.1 summarizes the advantages and drawbacks of the different approaches. Next, we review a number of specific approaches to learning a skill at the symbolic and trajectory levels.

Symbolic approaches

A large body of work uses a symbolic representation of both the learning and the encoding of skills and tasks (Muench et al., 1994; Friedrich et al., 1996;

Hovland et al., 1996; Nicolescu & Mataric, 2003; Pardowitz et al., 2007; Ekvall & Kragic, 2006; Saunders et al., 2006; Alissandrakis, Nehaniv, & Dautenhahn, 2007). This symbolic encoding may take on several forms. One common way is to segment and encode the task according to sequences of *predefined* actions described symbolically. Encoding and regenerating the sequences of these actions can then be done using classical machine learning techniques, such as HMM, see e.g., Hovland et al. (1996).

Often, these actions are encoded in a hierarchical manner. Nicolescu and Mataric (2003) employed a graph-based approach to generalize an object's transporting skill by using a wheeled mobile robot. In the model, each node in the graph represents a complete behavior and the generalization takes place at the level of the topological representation of the graph. Pardowitz et al. (2007) and Ekvall and Kragic (2006) followed a similar hierarchical and incremental approach to encode various household tasks (such as setting the table, see Fig. 1.9). In this case, learning consists of extracting symbolic rules managing

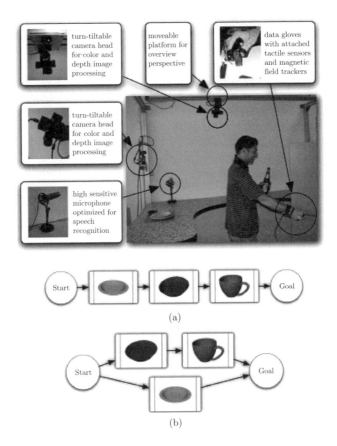

Figure 1.9 *Top:* A training center with dedicated sensors. *Bottom:* Precedence graphs learned by the system for a *"setting the table"* task. (a) The initial task precedence graph for the first three demonstrations. (b) The final task precedence graph after observing additional examples. Adapted from Pardowitz et al. (2007).

the way each object must be handled. Saunders et al. (2006) also exploited a hierarchical approach to encoding a skill in terms of pre-defined behaviors. The skill consisted in moving through a maze where a wheeled robot must avoid several kinds of obstacles and reach a set of specific subgoals. The novelty of the approach was that it used a symbolic representation of the skill to explore the teacher's role in guiding the incremental learning of the robot. Alissandrakis et al. (2007) also selected a symbolic approach for encoding human motions as sets of pre-defined postures, positions or configurations and considered various levels of granularity for the symbolic representation of the motion. This *a priori* knowledge was then utilized to explore correspondence problems through several simulated setups, including motion of articulated links in joint space and displacements of objects on a 2D plane (Fig. 1.6).

To summarize, the main advantage of this approach resides in an efficient learning of high-level skills, consisting of sequences of symbolic cues, through an interactive process. The drawback is that the associated methods rely on a large amount of prior knowledge to predefine the symbolic representation and to segment these efficiently (see Table 1.1).

Trajectory-based approaches

A proper choice of the variables when encoding a particular movement is crucial as it provides part of the solution to the problem of defining what is important to imitate. Work in PbD encodes human movements in either joint space, task space or torque space (Ude, 1993; Yang, Xu, & Chen, 1997; Yamane & Nakamura, 2003). The encoding may be specific to a cyclic motion (Ito, Noda, Hoshino, & Tani, 2006), a discrete motion (Calinon, Guenter, & Billard, 2007), or to a combination of both (Ijspeert, Nakanishi, & Schaal, 2003).

Encoding often encompasses the use of dimensionality reduction techniques that project the recorded signals into a latent space of motion of reduced dimensionality. The use of probabilistic models is appealing due to their generalization capabilities and to the compact encoding of the skills. Promising approaches when it comes to encoding human movements also include those that encapsulate the dynamics of the movement into the encoding itself (Bullock & Grossberg, 1989; Mussa-Ivaldi, 1997; Li & Horowitz, 1999; Schoener & Santos, 2001; Ijspeert et al., 2003). These different methods are highlighted below.

Skill encoding based on latent space projection

One issue in motor learning involves the difficulty of learning highly dimensional motor signals. To deal with this problem, research efforts have focused on trying to represent the high-dimensional motor space in a subspace of lower dimensionality. In the search for a latent space encapsulating the main characteristics of the motion, several linear and non-linear methods have been proposed. Among the linear methods, *Principal Component Analysis* (PCA) is one of the most common and is used in various fields of research where the datasets comprise human motion data (Chalodhorn, Grimes, Maganis, Rao, & Asada, 2006; Urtasun, Glardon, Boulic, Thalmann, & Fua, 2004). This simple technique is attractive since it is computationally fast and it remains generic for a wide range of signals, thus permitting its use as a pre-processing step to reduce the dimensionality for further analysis. Another body of work focuses on local dimensionality reduction techniques to deal with the non-linearities of the motion in latent space (Vijayakumar & Schaal, 2000; Vijayakumar, D'souza, & Schaal, 2005; Kambhatla, 1996). Global non-linear methods have also been proposed for the analysis of human motion (for transferring a skill to a humanoid robot) by a *Gaussian Process Latent Variable Model* (GPLVM) (Shon, Grochow, & Rao, 2005; Grochow, Martin, Hertzmann, & Popovic, 2004; Lawrence, 2005), *Isomap* (Jenkins & Mataric, 2004), or *Nonlinear Principal Component Neural Networks* (NLPCNN) (MacDorman, Chalodhorn, & Asada, 2004).

Skill encoding based on statistical modeling
An important aspect when transferring a skill to a robot is to concentrate on the ability of the robot to recognize a set of gestures that is further used to control the robot and/or teach new skills based on previously learned gestures. Various methods have been proposed to permit the recognition of gestures within a human-robot interaction framework. Brethes, Menezes, Lerasle, and Hayet (2004) suggested the use of *particle filters* to track the position and posture of hands using a vision system. Ardizzone, Chella, and Pirrone (2000) utilized *Support Vector Machine* (SVM) to recognize simple upper-body postures. Zoellner, Rogalla, Dillmann, and Zoellner (2002) employed SVM to recognize dynamic hand grasps through the use of a glove-based input device. Waldherr, Romero, and Thrun (2000) used a feed-forward neural network to recognize gestures for the control of a wheeled robot.

A number of authors have exploited the robustness of *Hidden Markov Models* (HMMs) to encode the temporal and spatial variations of complex signals. With a glove-based interface within an HRI application, C. Lee and Xu (1996) employed HMMs to recognize hand gestures from the sign language alphabet. Nam and Wohn (1996) utilized vision-based HMMs to recognize hand posture and hand motion. Park et al. (2005) also employed vision-based HMMs to recognize upper-body gestures used to control a humanoid robot. Fujie, Ejiri, Nakajima, Matsusaka, and Kobayashi (2004) and Kapoor and Picard (2001) used HMMs for a conversational robot to recognize para-linguistic information such as head nods and head shakes, which could then clarify whether the user's knowledge was correctly transferred to the robot.

In the applications mentioned above, HMMs are mainly used for recognition purposes. One trend of research investigates how HMMs can also deal with the high variability inherent to the demonstrations for reproduction purposes. In Computer Graphics, Brand and Hertzmann (2000) suggested to employ HMMs to identify common elements in a motion and synthesize new motions by combining and blending these elements. Tso and Liu (1996) used HMM to encode and retrieve Cartesian trajectories, where one of the trajectories contained in the training set could reproduce the skill (by keeping the trajectory of the dataset with the highest likelihood, i.e., the one that generalized the most as compared to the others). Yang et al. employed HMMs to encode the motion

of a robot's gripper, in the *joint space* or in the *task space*, by considering either
its positions or its velocities (Yang et al., 1997).

The *Mimesis Model* (Inamura, Toshima, & Nakamura, 2003; Inamura
et al., 2005; D. Lee & Nakamura, 2006, 2007; Kulic, Takano, & Nakamura,
2008) follows an approach in which the HMM encodes a set of trajectories,
and where multiple HMMs can be used to retrieve new generalized motions
through a stochastic process. This method renders it possible to transfer human
motion to a humanoid robot and to blend various motions (Takano, Yamane,
& Nakamura, 2007). However, the averaging process used for reproduction
does not guarantee that the system retrieves smooth trajectories. Alternative
approaches have been proposed to pre-decompose the trajectories into a set
of relevant keypoints and to retrieve a generalized version of the trajectories
through spline fitting techniques (Calinon, Guenter, & Billard, 2005; Asfour,
Gyarfas, Azad, & Dillmann, 2006; Aleotti & Caselli, 2006b).

Another approach based on HMM, originally used for speech recognition and speech synthesis, involves an augmention of the observation vectors
of the HMM by the differences of nearby observations, called δ *features*, where
the relationships between the static and dynamic features are explicit (Furui,
1986). However, these features are originally considered as being independent when modeled by the HMM, which results in inconsistencies between
static and dynamic characteristics. Recently, Zen, Tokuda, and Kitamura
(2007) re-formulated the problem by imposing explicit relationships between
the static and dynamic features and proposed a statistically correct model
called *trajectory-HMM* for speech recognition and speech synthesis. This model
can be trained by dedicated methods proposed by the authors. A first attempt
at extending this framework to a robot learning by imitation was proposed by
Sugiura and Iwahashi (2008).

Skill encoding based on dynamical systems
Dynamical systems offer a particularly interesting solution to imitation processes aimed at being robust to perturbations (Yamane & Nakamura, 2003;
Nakanishi et al., 2004; Righetti & Ijspeert, 2006). One of the first studies emphasizing this approach in learning by imitation was that of Ijspeert
et al. (2003), where a motor representation based on dynamical systems was
employed for encoding movements and for replaying them under various conditions. The approach combines two ingredients: non-linear dynamical systems
for the robust encoding of the trajectories, and techniques from non-parametric
regression for shaping the attractor landscapes according to the demonstrated
trajectories. The essence of the approach resided in starting with a simple
dynamical system, e.g., a set of linear differential equations, and transforming it into a non-linear system with prescribed attractor dynamics by means
of a learnable autonomous forcing term. With this method, one can generate
both point attractors and limit cycle attractors of almost arbitrary complexity.
Both types of attractors have been successfully used to encode discrete (e.g.,
reaching) and rhythmic movements (e.g., drumming).

Locally Weighted regression (LWR) has initially been proposed to permit
the learning of the above system parameters (Atkeson, 1990; Atkeson, Moore,

& Schaal, 1997; Moore, 1992; Ghahramani & Jordan, 1994). It can be viewed as a memory-based method combining the simplicity of linear least squares regression with the flexibility of non-linear regression. Further work has mainly been concentrated on moving on from a memory-based approach to a model-based one, as well as progressing from a batch learning process to an incremental learning strategy (Schaal & Atkeson, 1996, 1998; Vijayakumar et al., 2005). Schaal and Atkeson (1998) use *Receptive Field Weighted Regression* (RFWR) as a non-parametric approach to incrementally learn the fitting function without the need for storing the whole training data in memory. Vijayakumar et al. (2005) suggested that one use *Locally Weighted Projection Regression* (LWPR) to improve the means of operating efficiently in high dimensional space.

Ito et al. (2006) proposed an alternative way to implicitly encode the dynamics of bimanual and unimanual tasks, based on recurrent neural networks. It is of interest that the network can switch between motions in a smooth manner when separately trained to encode two sensorimotor loops. The approach has been validated for modeling dancing motions during which the user first initiates imitation and where the roles of the imitator and demonstrator can be dynamically interchanged. They also demonstrate the ability for cyclic manipulation tasks to be learned, in which the robot continuously moves a ball from one hand to the other and is able to switch dynamically to another cyclic motion consisting in lifting and releasing the ball.

Another approach, investigated by Mayer et al. (2007), involves using principles known from fluid dynamics to reproduce a skill at a trajectory level in a PbD context. Their approach considers multiple demonstrations of a skill as a basis for the reproduction and generalization of the skill to various contexts (different initial positions or the presence of new obstacles) by computing new trajectories through simulation. The generalization is performed by considering the trajectory as a fluid that adapts itself to new situations by bypassing obstacles and by smoothing out discontinuities. The system is put to the test with a knot-tying task in surgery. A drawback of applying this method to PbD is the length of the generalization process (more than 10 minutes). It can, thus, hardly be used for interactive teaching interactions.

1.4.3   Incremental teaching methods

We have seen that PbD systems should be capable of learning a task from as few demonstrations as possible. The robot should thus be able to start performing its tasks right away, and gradually improve its performance while at the same time being monitored by the user. Incremental learning strategies that gradually refine the task knowledge as more examples become available pave the way towards PbD systems suitable for real-time human-robot interactions.

These incremental learning methods employ various forms of deixis, both verbal and non-verbal, to guide the robot's attention to the important parts of the demonstration or to particular mistakes it produces during the reproduction of the task. Such incremental and guided learning is often referred to as the *scaffolding* or *molding* of a robot's knowledge. It is deemed most important to allow the robot to learn tasks of increasing complexity (Breazeal, Berlin, Brooks, Gray, & Thomaz, 2006; Saunders et al., 2006).

The application of incremental learning techniques to PbD has contributed to the development of methods for learning complex tasks within the household domain from as few demonstrations as possible. It has also contributed to the development and application of machine learning, allowing continuous and incremental refinement of the task model. Such systems have sometimes been referred to as *background-knowledge-based* or *EM-deductive PbD systems*, as presented in (Sato, Bernardin, Kimura, & Ikeuchi, 2002; Zoellner, Pardowitz, Knoop, & Dillmann, 2005). They usually require very few or even only a single demonstration in order to generate executable task descriptions. The main objective of this type of work is to build a meta-representation of the knowledge required by the robot on the task and to apply reasoning methods to this knowledge database. Such reasoning involves recognizing, learning and representing repetitive tasks. Pardowitz, Zoellner, and Dillmann (2005) have discussed how different forms of knowledge can be balanced in an incremental learning system and have proposed a system that relies on building *task precedence graphs*. Such graphs encode hypotheses that the system makes on the sequential structure of a task. By learning the task precedence graphs, the system can schedule its operations most flexibly while still meeting the goals of the task, where a temporal precedence relation can be learned incrementally (see Fig. 1.9).

1.4.4   Human-robot interaction in PbD

Another perspective adopted by PbD in order to render the transfer of skill more efficient is to focus on the interaction aspect of the transfer process. As this transfer problem is complex and involves a combination of social mechanisms, several insights from *Human-Robot Interaction* (HRI) have been explored to increase the efficiency use of the teaching capabilities of the human user.

The development of algorithms for detecting *social cues* (given implicitly or explicitly by the teacher during training) and their integration as part of other generic mechanisms for PbD has become the focus of a large body of work. Such social cues can be viewed as a way to introduce priors within a statistical learning system, and, by so doing, speed up learning. Indeed, several hints can be used to transfer a skill, not only the demonstration of a task multiple times but also the highlighting of important components of the skill. This can be achieved by various means and through several modalities.

A large body of work has been dedicated to the use of pointing and gazing as a way of conveying the intention of the user (Scassellati, 1999; Kozima & Yano, 2001; Ishiguro, Ono, Imai, & Kanda, 2003; Nickel & Stiefelhagen, 2003; Ito & Tani, 2004; Hafner & Kaplan, 2005; Dillmann, 2004; Breazeal, Buchsbaum, Gray, Gatenby, & Blumberg, 2005). Vocal deixis, using standard speech recognition engines, has also been widely explored (Dominey et al., 2005; Dillmann, 2004; Pardowitz et al., 2007). Thomaz, Berlin, and Breazeal (2005) and Breazeal and Aryananda (2002) considered only the prosody of the speech pattern, rather than the exact content of the speech, as a way of inferring

some information on the user's communicative intent. In Calinon and Billard
(2006), these social cues are learned through an imitative game, whereby the
user imitates the robot. This allows the robot to build a user-specific model of
the social pointers, and to become more robust at detecting them.

Finally, a core idea of the HRI approach to PbD is that imitation is
goal-directed. In order words, actions are meant to fulfill a specific purpose
and convey the intention of the actor (Bekkering et al., 2000). While a longstanding trend in PbD involves approaching the problem from the standpoint
of trajectory following (Nicolescu & Mataric, 2007; J. Demiris & Hayes, 1996;
Gaussier, Moga, Banquet, & Quoy, 1998) and joint motion replication (Aleotti
& Caselli, 2006b; Ogino, Toichi, Yoshikawa, & Asada, 2006; Billard & Matarić,
2001; Ijspeert, Nakanishi, & Schaal, 2002a), recent work, inspired by the above
rationale, starts from the assumption that imitation is not merely about observing and replicating the motion, but rather about *understanding* the goals of a
given action. The ability to learn to imitate relies on the imitator's capacity to
infer the demonstrator's intentions (Erlhagen et al., 2006; Breazeal et al., 2006).
However, demonstrations may be ambiguous and extracting the intention of the
demonstrator requires the building of a cognitive model of him/her (Jansen &
Belpaeme, 2006; Chella, Dindo, & Infantino, 2006), as well as exploiting other
social cues to provide complementary knowledge (Breazeal et al., 2006; Saunders et al., 2006; Calinon & Billard, 2007b).

The study of the way humans learn both to extract the goals of a set of
observed actions and to give these goals a hierarchy of preference is fundamental
to the understanding of the underlying decisional process of imitation. Recent
work tackling these issues has followed a probabilistic approach to explain a
goal derivation and a sequential application. The explanation, in turn, renders it possible to learn manipulatory tasks that require the sequencing of a
goal subsets (Cuijpers, van Schie, Koppen, Erlhagen, & Bekkering, 2006; Pardowitz et al., 2005; Nicolescu & Mataric, 2007; Hoffman, Grimes, Shon, & Rao,
2006). Understanding the goal of the task is still only half of the picture as
there may be several ways of achieving a certain goal. Moreover, what suits
the demonstrator may not necessarily be suitable for the imitator (Nehaniv
& Dautenhahn, 2000). Thus, different models may be allowed to compete to
obtain a solution that is optimal both from the point of view of the imitator
and of the demonstrator (Y. Demiris & Khadhouri, 2006; Billard, Calinon, &
Guenter, 2006).

Figure 1.10 presents a robotic experiment that replicates psychological
experiments conducted with young children as proposed in Bekkering et al.
(2000). Children have a tendency to substitute ipsilateral gestures for contralateral ones when dots are present. In contrast, when dots are absent from the
demonstration, the number of substitutions drops significantly. Thus, despite
the fact that the gesture is the same in both conditions, the presence or absence
of a physical object (the dot) has a dramatic effect on the reproduction. When
the object is present, object selection takes the highest priority. Children, then,
nearly always direct their imitation to the appropriate target object, at the cost
of selecting the "wrong" hand. When removing the dots, the complexity of the
task (the number of constraints to satisfy) decreases, and, hence, constraints

"Dots" condition   "No dots" condition

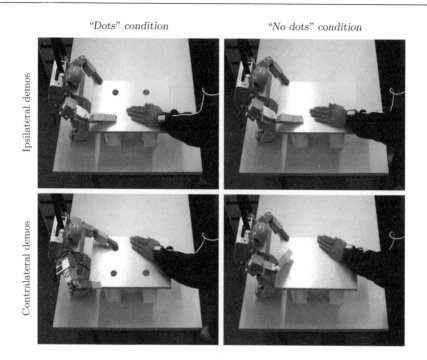

Figure 1.10 Goal-directed experiments with robots proposed in Calinon et al. (2005) based on the replication of psychological experiments conducted with young children (Bekkering et al., 2000). In this experiment with a humanoid robot, the trajectory of the demonstrator's arm and hand path might not permit the unequivocal determination of the joint angle trajectories of the robot's arm. Indeed, depending on where the target is located, several constraints (goals) might compete, and satisfying all of them would not always lead to a solution. In the case of a contralateral motion, the robot's arm is too small to be able to both reach the target and perform the same gesture. In the presence of dots, the robot needs to find a trade-off between satisfying each of the constraints. In this situation, it selects an ipsilateral motion, and, thus, prioritizes the constraint of reaching for the object. This amounts to determining the importance of each constraint with respect to one another.

of lower importance can be fulfilled (such as producing the same gesture or using the same hand). The experiment starts with the (human) demonstrator and the (robot) imitator standing in front of a table, facing each other. On both sides of the table, two colored dots have been stamped at equal distance to the demonstrator and imitator's starting positions. In a first set of demonstrations, the demonstrator reaches for each dot alternatively with his left and right arm. If the demonstrator reaches for the dot on the left-hand side of the table with his/her left arm, he/she is said to perform an ipsilateral motion. If conversely the demonstrator reaches for the dot on the right-hand side of the table with his/her left arm, he/she is said to perform a contralateral motion. Subsequently, the demonstrator produces the same ipsilateral and contralateral motions, but without the presence of dots. Each of these motions

are demonstrated five consecutive times. In each case, the demonstrator starts
from a similar starting position. While observing the demonstration, the robot
tries to make sense of the experiment and extracts the demonstrator's intention underlying the task by determining a set of constraints for the task in a
statistical manner. When the demonstration ends, the robot computes the trajectory that best satisfies the constraints extracted during the demonstration
and generates a motion that follows this trajectory.

1.4.5   Biologically-oriented learning approaches

Another important trend in robot PbD takes a more biological stance and
develops computational models of imitation learning in animals. We here briefly
review recent progress in this area.

Conceptual Models of Imitation Learning
Bioinspiration is first revealed in the conceptual schematic of the sensorimotor
flow that is at the basis of imitation learning, followed by various authors in
PbD over the years.
    Visual sensory information needs to be parsed into information relating to
objects and their spatial location in an internal or external coordinate system;
the depicted organization is largely inspired by the dorsal (what) and ventral
(where) stream as discovered in neuroscientific research (Maunsell & Van Essen,
1983). As a result, the posture of the teacher and/or the position of the object
while moving (if one is involved) should become available. Subsequently, one
of the major questions revolves around how such information can be converted
into action through the concept of movement primitives, also called "movement
schemas", "basis behaviors", "units of action", or "macro actions" (Arbib,
Iberall, & Lyons, 1985; Nehaniv, 2003; Sternad & Schaal, 1999; Sutton, Singh,
Precup, & Ravindran, 1999). Movement primitives are sequences of actions
accomplishing a complete goal-directed behavior. They can be as simple as
elementary actions of an actuator (e.g., "go forward", "go backwards", etc.),
but, as discussed in Schaal (1999a), such low-level representations do not scale
well to learning in systems with many degrees of freedom. Thus, it is useful
for a movement primitive to code complete temporal behaviors, like "grasping
a cup", "walking", "a tennis serve", etc. Y. Demiris and Hayes (2002) and
Wolpert and Kawato (1998) suggested that the perceived action of the teacher
is mapped onto a set of existing primitives in an assimilation phase. This
mapping process also needs to resolve the correspondence problem concerning
a mismatch between the teacher's and the student's bodies. Subsequently,
one can adjust the most appropriate primitives by learning to improve the
performance in an accommodation phase. If no existing primitive is a good
match for the observed behavior, a new primitive must be generated. After an
initial imitation phase, self-improvement, e.g., with the help of a reinforcement-
based performance evaluation criterion (Sutton & Barto, 1998), can refine both
movement primitives and an assumed stage of motor command generation until
a desired level of motor performance is achieved.

Neural Models of Imitation Learning

Bioinspiration is also revealed in the development of neural network models of the mechanisms at the basis of imitation learning. Current models of Imitation Learning all ground their approach on the idea that imitation is at core driven by a *mirror neuron* system. The mirror neuron system refers to a network of brain areas in premotor and parietal cortices which is activated by both the recognition and the production of the same kind of object-oriented movements performed by oneself and by others. The reader is referred to (Rizzolatti et al., 2001; Decety et al., 2002; Iacoboni et al., 1999) for recent reports on this system in monkeys and humans, in addition to its link to imitation.

Models of the mirror neuron system assume that, somewhere, sensory information concerning the motion of others as well as self-generated motion is coded using the same representation and in a common brain area. The models, however, differ in the way they represent this common center of information. Besides, while all models draw from the evidence of the existence of a mirror neuron circuit and of its application to explain multimodal sensory-motor processing, they, nevertheless, go further and tackle the issue of how the brain manages to process the flow of sensorimotor information that is at the basis of how one observes and produces actions. For a comprehensive and critical review of computational models of the mirror neuron system, the reader is referred to Oztop, Kawato, and Arbib (2006). Next, we briefly review the most recent research efforts among these authors.

One of the first approaches to explaining how the brain processes visuo-motor control and imitation is based on the idea that it relies on Forward Models to compare, predict and generate motion (Wolpert & Kawato, 1998). This early work sets the ground for some of the current neural models of the visuo-motor pathway underlying imitation. For instance, a recent study by Demiris and colleagues combined the evidence that there is a mirror neuron system constituting the basis of recognition and production of basic grasping motions and which proves the existence of forward models for guiding these motions (Y. Demiris & Hayes, 2002; Y. Demiris & Khadhouri, 2006). The models contribute both to the understanding of the mirror neuron system (MNS) in animals as well as to its use in controlling robots. For instance, the models successfully reproduce the timing of neural activity in animals while observing various grasping motions and recreating the kinematics of arm motions during these movements. The models have been implemented to control reaching and grasping motions in both humanoid and non-humanoid robots.

In contrast, Arbib and colleagues have taken a strong biological stance by following an evolutionary approach to modeling the Mirror Neuron System (MNS), emphasizing its application to robotics in merely a second stage. They hypothesized that the ability to imitate found in humans has evolved from the ability of monkeys to reach and grasp and from the ability of chimpanzees to perform simple imitation. They developed models of each of these evolutionary stages, starting from a detailed one of the neural substrate underlying the ability of monkeys to reach and grasp (Fagg & Arbib, 1998). The model was subsequently extended to include the monkey MNS (Oztop & Arbib, 2002) after which the authors finally moved on to models of the neural circuits underlying

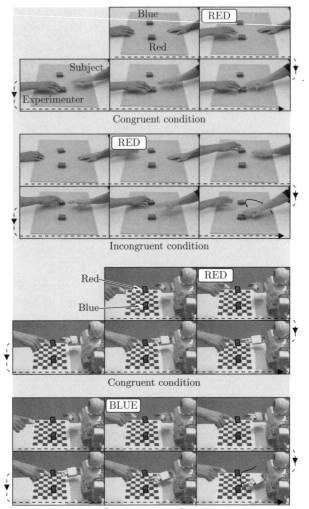

Congruent condition

Incongruent condition

Congruent condition

Incongruent condition

Figure 1.11 A human experimenter, a human imitator and a robot imitator playing a simple imitation game in which the imitator must point to the object named by the experimenter and not to the object which the experimenter points at. The robot's decision process is controlled by a neural model similar to that found in humans. As a result, it experiences the same associated deficit, known as the principle of *ideomotor compatibility*, stating that *observing the movements of others influences the quality of one's own performance.* When presented with conflicting cues (incongruent conditions), e.g., when the experimenter points at a different object than the one named, the robot, like the human subject, either fails to reach for the correct object, or hesitates and later corrects the movement. Adapted from Sauser and Billard (2006a).

the human's ability to imitate (Arbib, Billard, Iacoboni, & Oztop, 2000). The models replicate findings of brain imaging and cell recording studies, and further make predictions on the time course of the neural activity for motion prediction. As such, they reconciliate a view of forward models of action with an immediate MNS representation of these same actions. They have been used to control reaching and grasping motions in humanoid robotic arms and hands (Oztop, Lin, Kawato, & Cheng, 2006).

Expanding on the concept of an MNS for sensorimotor coupling, Sauser & Billard explored the use of competitive neural fields to explain the dynamics underlying a multimodal representation of sensory information and the way the brain may disambiguate and select across competitive sensory stimuli to proceed to a given motor program (Sauser & Billard, 2006c, 2006b). This work explains the principle of *ideomotor compatibility*, by which observing the movements of others influences the quality of one's own performance. Consequently, neural models accounting for a set of related behavioral studies have been developed (Brass, Bekkering, Wohlschlaeger, & Prinz, 2001). These models expand the basic mirror neuron circuit to explain the consecutive stages of sensory-sensory and sensory-motor processing at the basis of this phenomenon. Sauser and Billard (2006a) discuss how the capacity for ideomotor facilitation can provide a robot with human-like behavior at the expense of several disadvantages such as hesitation and even mistakes (see Fig. 1.11).

CHAPTER 2

SYSTEM ARCHITECTURE

This chapter presents the various models and algorithms used in the proposed PbD probabilistic framework. We first treat encoding and reproduction issues, and then discuss the issues of learning parameters, pre-processing steps and the regularization of parameters.[1] Most of the models and algorithms presented here are not new, however their use in a PbD framework has yet to be explored. This implies the adaptation and combination of the different methods to build a probabilistic system allowing to teach a humanoid robot new skills and to re-use the learned skill in other situations.

For an easier comprehension of the framework, the various algorithms presented in this chapter are illustrated by simple datasets of mostly one-dimensional trajectories.[2] Real-world applications using human motion data processed by the proposed models is provided in the following chapters.

2.1 ILLUSTRATION OF THE PROPOSED PROBABILISTIC APPROACH

First, Figures 2.1 and 2.2 present an example to motivate the continuous approach adopted throughout this book, that of extracting task constraints within a probabilistic framework. Let us consider a task for which the aim is to place a ball on the right-hand side of object B, and where object A is irrelevant for the task. In Figure 2.1, a qualitative description of the ball position relative to the two objects A and B is used to describe the task at a *symbolic level*. *"Right A"* and *"Left A"* correspond to a final position of the ball on the right-hand side and on the left-hand side of object A. Similarly, *"Right B"* and *"Left B"* correspond to a final position of the ball on the right-hand side and on the left-hand side of object B. The histogram plots show the extraction of task constraints through discrete statistics. After the first demonstration, the

---

[1] Note that the different algorithms are not presented in their processing order but in their order of importance for the comprehension of the complete system.

[2] Note that a corresponding set of *Matlab* codes written by the author for the algorithms presented in this chapter is available online at http://www.programming-by-demonstration.org/book.

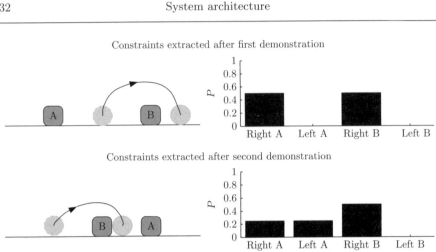

Figure 2.1 The extraction of task constraints at a symbolic level. The initial and final positions of the ball are respectively represented with a dashed and solid line.

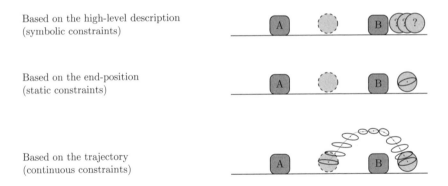

Figure 2.2 The extraction of task constraints by using various levels of represen-tation of the constraints. The initial and final positions of the ball are respectively represented with a dashed and solid line.

ball is positioned on the right-hand side of the two objects. The probabilities that these two constraints are relevant thus become equal ($\mathcal{P} = 0.5$). After the second demonstration, performed for a new initial condition, the ball is moved to the left-hand side of object A and to the right-hand side of object B. The probability that this last event is relevant for the task then becomes higher than for the other events. After two demonstrations, the principal constraint of the task detected is thus to place the ball on the right-hand side of object B. This constraint has been extracted from multiple observations of the same task performed in various initial configurations.

Figure 2.2 shows the extraction of the task constraints by several levels of representation. When the constraints are based solely on a symbolic descrip-tion, the system can fail to reproduce a coherent behavior due to there not existing information concerning the final distance of the ball with respect to object B (*first row*). For instance, if object B represents a plate and the ball represents a knife, it would be awkward to place the knife one meter away from the plate when setting a table. Regardless of whether the knife is on the right-hand side of the plate, qualitatively speaking, the task cannot be said to be correctly reproduced. This example shows that the simple *left-hand side* and *right-hand side* rules defined to extract the symbolic features of the task are inappropriate in certain situations for a generic reproduction of the skill.

A constraint based on the final relative position of the objects would be more representative (Fig. 2.2, *second row*). By defining an average position of the ball at the end of the motion as well as a covariance matrix determining the variations allowed around this position, the problem induced by the symbolic description of the task is partly resolved. The center, μ, is now used to describe the final relative position of the ball with respect to object B. The covariance matrix, Σ, (represented by an ellipse) describes a permissible area where the object should be placed, defined by the distribution probabilities over a given threshold (the contour of the ellipse represents one standard deviation of the normal distribution).

However, with this static representation of the task constraints, there is still a lack of information concerning the way the object is moved, which can also be relevant for the task. For example, the two demonstrations show that the motion should follow a bell-shaped trajectory in order to displace the ball without hitting the other objects. This particularity of the task can be learned by considering constraints at a trajectory level (Fig. 2.2, *third row*). By extending the representation defined in the *second row* to the positions at each time step along the movement, a temporally continuous representation of the task constraints can be considered, defined by an averaged trajectory and associated covariance matrices, represented at each time step by the mean position of the ball and the variations/correlations allowed around this center.

To this end, a generic methodology is proposed to encode temporally continuous gestures through a probabilistic representation based on a *Gaussian Mixture Model* (GMM), which allows the autonomous and incremental construction of a model of the skill. Having a probabilistic model of a skill renders it possible: (1) to provide a compact and efficient representation of high-dimensional data; (2) to classify existing and detect new skills; (3) to evaluate a reproduction attempt; (4) to extract the constraints of the skill; and (5) to reproduce the learned skill in a different situation by generalizing over multiple observations. We suggest to extract the constraints of the task in a probabilistic and continuous form through *Gaussian Mixture Regression* (GMR), which is also employed to define a metric of imitation performance and to determine an optimal controller for the robot.

The term *constraint* is used throughout this book to define the behavior of a set of data collected through multiple demonstrations of a skill. The constraints are considered at a trajectory level by computing a local estimation of the variations and correlations allowed across the different variables representing the task. Thus, the constraints are represented by a set of centers and covariance matrices evaluated at several time intervals along the movement.

This process aims at finding a flexible controller to reproduce the task in a variety of situations, in order for the system to retrieve smooth generalized motions that can be easily combined and run on the robot in situations that have not been explicitly demonstrated.

2.2 ENCODING OF MOTION IN A *GAUSSIAN MIXTURE MODEL* (GMM)

A dataset composed of N datapoints of D dimension is considered. For a *Gaussian Mixture Model* (GMM) of K Gaussian distributions, the probability that one of the datapoints ξ_j belongs to the GMM is defined by

$$\mathcal{P}(\xi_j) = \sum_{k=1}^{K} \mathcal{P}(k)\,\mathcal{P}(\xi_j|k), \qquad (2.1)$$

where $\mathcal{P}(k)$ is a prior probability and $\mathcal{P}(\xi_j|k)$ is a conditional probability density function defined by the parameters

$$\mathcal{P}(k) = \pi_k,$$

$$\mathcal{P}(\xi_j|k) = \mathcal{N}(\xi_j; \mu_k, \Sigma_k) = \frac{1}{\sqrt{(2\pi)^D|\Sigma_k|}}\, e^{-\frac{1}{2}\left((\xi_j-\mu_k)^\top \Sigma_k^{-1}(\xi_j-\mu_k)\right)}.$$

The GMM is then described by the set of parameters $\{\pi_k, \mu_k, \Sigma_k\}_{k=1}^{K}$, respectively representing the *prior* probabilities, *centers* and *covariance* matrices of the model. The *prior* probabilities, π_k, satisfy $\pi_k \in [0,1]$ and $\sum_{k=1}^{K}\pi_k = 1$. We first consider the parameters of the GMM to be known (later, we see in Section 2.6.1 how these are learned). Figure 2.3 illustrates the encoding of a set of datapoints in a GMM.

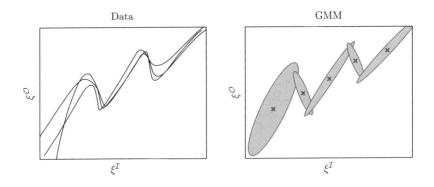

Figure 2.3 A dataset of $D = 2$ dimensions encoded in a GMM of $K = 5$ components, where the Gaussian distributions are used to model $\xi = [\xi^\mathcal{I}, \xi^\mathcal{O}]$ consisting of three demonstrations of trajectories.

Throughout this work, the notation $\mathcal{N}(\mu, \Sigma)$ is used to define a Gaussian distribution and $\mathcal{N}(\xi_j; \mu, \Sigma)$ to define the probability of a datapoint ξ_j belonging to this distribution.

2.2.1   Recognition, classification and evaluation of a reproduction attempt

A GMM can be used to recognize gestures by estimating the likelihood of the observed data having been generated by the model. The log-likelihood of a GMM defined by parameters $\{\pi_k, \mu_k, \Sigma_k\}_{k=1}^{K}$ when testing a set of N datapoints $\xi = \{\xi_j\}_{j=1}^{N}$, is defined by

$$\mathcal{L}(\xi) = \sum_{j=1}^{N} \log\left(\mathcal{P}(\xi_j)\right), \tag{2.2}$$

where $\mathcal{P}(\xi_j)$ is the probability of the datapoint ξ_j being generated by the model defined in (2.1). This log-likelihood can similarly be used to evaluate a reproduction attempt.

2.3   ENCODING OF MOTION IN *HIDDEN MARKOV MODEL* (HMM)

Similarly to GMM, the *Hidden Markov Model* (HMM), in its continuous form, utilizes a mixture of multivariate Gaussians to describe the distribution of a dataset (Rabiner, 1989). The main difference resides in HMM also having the ability to encapsulate transition probabilities between the sets of Gaussian distributions, offering a method to probabilistically describe the temporal variations in the dataset. HMM can thus be seen as a double stochastic process, based on transition probabilities, output distributions and initial state distributions. We principally consider the case where each state is described by a single multivariate Gaussian distribution, $\mathcal{N}(\mu_i, \Sigma_i)$. Thus, an HMM can be represented by considering each component i of the GMM as an output distribution for the corresponding state i in the HMM. Π_i is the probability of starting from state i, and a_{ij} is the probability of transiting from state i to state j. The complete set of parameters designating an HMM is thus defined by $\{\Pi_i, a_{ij}, \mu_i, \Sigma_i\}$.

Figure 2.4 and Table 2.1 present an example of an HMM encoding of 3 demonstrations of 2-dimensional trajectories. In this example, the considered model is fully-connected (each state is connected to every other state), with an initialization of the parameters biased toward a left-right topology. In this way, we remain generic concerning the architecture of the model but we bias the solution toward the one expected from the considered dataset. In this example, due to the dataset in question not showing periodic nor recurrent behavior, the parameters naturally converge to a *left-right* topology, i.e., some of the transition parameters converge to zero (learning of the parameters is

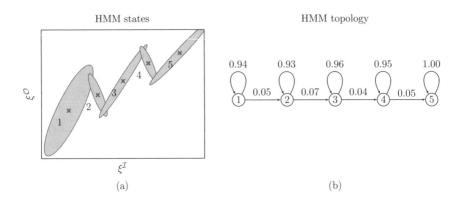

(a)   (b)

Figure 2.4 A dataset of $D = 2$ dimensions encoded in an HMM of 5 states used for modeling $\mathcal{P}(\xi^{\mathcal{I}}, \xi^{\mathcal{O}})$. The dataset consists of three demonstrations of trajectories, where each trajectory is defined by two variables, $\xi^{\mathcal{I}}$ and $\xi^{\mathcal{O}}$. (a) Output distributions representing the data. (b) The final topology of the model and transition probabilities between the various states.

Table 2.1 Transition probabilities, a_{ij}, and initial state probabilities, Π_i, for a fully-connected HMM scheme encoding continuous data (see also Fig. 2.4).

Transition probabilities a_{ij}					
i \ j	1	2	3	4	5
1	**0.94**	**0.05**	0.00	0.00	0.00
2	0.00	**0.93**	**0.07**	0.00	0.00
3	0.00	0.00	**0.96**	**0.04**	0.00
4	0.00	0.00	0.00	**0.95**	**0.05**
5	0.00	0.00	0.00	0.00	**1.00**

Initial state probabilities Π_i				
1	2	3	4	5
1.00	0.00	0.00	0.00	0.00

further described in Sect. 2.6.2). Table 2.1 shows the transition probabilities of the model, with non-zero values for the recurrent transitions (probability of remaining in the same state) and for the transition from state i to state $i + 1$ (transition to the second closest Gaussian distribution).

HMM is an appealing model in PbD because of its capacity to locally model the properties of multidimensional human motion and since its probabilistic representation allows treatment of noisy real-world data. Indeed, the spatial variability of the data is probabilistically encapsulated through the Gaussian distributions, while the temporal variability is encapsulated through the transition probabilities between the states. As compared to GMM, HMM also renders it possible to deal with non-homogenous temporal deformation of the trajectories (and thereby act as a method to temporally align various trajectories).

2.3.1   Recognition, classification and evaluation of a reproduction attempt

The evaluation of an observation or a reproduction attempt $\xi = \{\xi_1, \xi_2, \ldots, \xi_T\}$ in HMM requires the use of a *forward-backward* procedure (Rabiner, 1989). Let us first define q_t as one of the K states underlying the observation at time t. The *forward* variable, $\alpha_{k,t}$, is defined as the probability of partially observing a sequence $\{\xi_1, \xi_2, \ldots, \xi_t\}$ of length t as well as of being in state k at time t, i.e., $\alpha_{k,t} = \mathcal{P}(\xi_1, \xi_2, \ldots, \xi_t, q_t = k)$. When parameters $\{\Pi, a, \mu, \Sigma\}$ of the HMM are known, and it can be assumed that each state of the HMM is represented by a single Gaussian distribution, $\alpha_{k,t}$ is inductively computed, starting from $\alpha_{k,1}$, as follows:

Initialization:

$$\alpha_{k,1} = \Pi_k \, \mathcal{N}(\xi_1; \mu_k, \Sigma_k) \qquad \Big| \; \forall k \in \{1, \ldots, K\}.$$

Induction:

$$\alpha_{k,t+1} = \left(\sum_{i=1}^{K} \alpha_{i,t} \, a_{ik} \right) \mathcal{N}(\xi_{t+1}; \mu_k, \Sigma_k) \qquad \begin{aligned} & \Big| \; \forall k \in \{1, \ldots, K\} \\ & \Big| \; \forall t \in \{1, \ldots, T-1\}. \end{aligned} \qquad (2.3)$$

Termination:

$$\mathcal{P}(\xi) = \sum_{i=1}^{K} \alpha_{i,T}.$$

The evaluation of a reproduction attempt $\xi = \{\xi_1, \xi_2, \ldots, \xi_T\}$ of length T is thus calculated by using the *forward* variable defined in (2.3) to compute $\mathcal{P}(\xi) = \sum_{i=1}^{K} \alpha_{i,T}$. This probability is usually represented in its log-likelihood form as in (2.2).

The *Viterbi* algorithm can be used to retrieve the sequence of states corresponding to an observed trajectory (Viterbi, 1967). Let us consider a sequence of states $\{q_1, q_2, \ldots, q_T\}$ and a sequence of observations $\{\xi_1, \xi_2, \ldots, \xi_T\}$ of length T (e.g., a 3D Cartesian path). We define the variable $\delta_{k,t}$ as the highest probability along a single sequence of states, at time step t, accounting for the first t observations and ending in state k, i.e., $\delta_{k,t} = \max \big[\mathcal{P}(q_1, q_2, \ldots, q_t = i, \xi_1, \xi_2, \ldots, \xi_t) \big]$, which can be computed by induction. The variable $\epsilon_{k,t}$ is also defined to keep track of the argument that maximizes $\delta_{k,t}$ for each time step t and state k. The optimal state sequences is defined by using the notation $\{\hat{q}_1, \hat{q}_2, \ldots, \hat{q}_T\}$. When parameters $\{\Pi, a, \mu, \Sigma\}$ of the HMM are known, $\delta_{k,t}$ can be recursively computed as follows:

Initialization:

$$\delta_{k,1} = \Pi_k \, \mathcal{N}(\xi_1; \mu_k, \Sigma_k) \qquad \Big| \; \forall k \in \{1, \ldots, K\},$$
$$\epsilon_{k,1} = 0 \qquad \Big| \; \forall k \in \{1, \ldots, K\}.$$

Induction:

$$\delta_{k,t} = \max_{i \in \{1, \ldots, K\}} (\delta_{i,t-1} \, a_{ik}) \; \mathcal{N}(\xi_t; \mu_k, \Sigma_k) \qquad \begin{aligned} & \Big| \; \forall k \in \{1, \ldots, K\} \\ & \Big| \; \forall t \in \{2, \ldots, T\}, \end{aligned}$$

$$\epsilon_{k,t} = \arg\max_{i \in \{1, \ldots, K\}} (\delta_{i,t-1} \, a_{ik}) \qquad \begin{aligned} & \Big| \; \forall k \in \{1, \ldots, K\} \\ & \Big| \; \forall t \in \{2, \ldots, T\}. \end{aligned}$$

Termination:

$$\hat{q}_T = \underset{i\in\{1,\ldots,K\}}{\arg\max} (\delta_{i,T}).$$

States sequence backtracking:

$$\hat{q}_t = \epsilon_{\hat{q}_{t+1},t+1} \qquad\qquad\qquad \big| \; \forall t \in \{T-1, T-2, \ldots, 1\}.$$

The underlying sequence of states corresponding to the observed trajectory ξ is then defined by $\hat{q}$.

2.4   REPRODUCTION THROUGH *GAUSSIAN MIXTURE REGRESSION* (GMR)

This section addresses the *Gaussian Mixture Regression* (GMR) process used to reproduce a learned skill, which is based on the theorem of Gaussian conditioning and on the linear combination properties of Gaussian distributions. We first recall in Summary Box 2.1 and Figure 2.5 certain properties of multivariate normal distributions that will be relevant for the description of the regression process. The complete proofs can be found in Bishop (2006), Mendel (1995) and Muirhead (1982).

From a set of trajectories described by a series of datapoints $\xi = [\xi^{\mathcal{I}}, \xi^{\mathcal{O}}]$ consisting of input and output components, we first model the joint probability distribution $\mathcal{P}(\xi^{\mathcal{I}}, \xi^{\mathcal{O}})$ in a GMM. The retrieval process then consists of assessing $\mathbb{E}\left[\mathcal{P}(\xi^{\mathcal{O}}|\xi^{\mathcal{I}})\right]$, with associated constraints estimated with $\text{cov}\left(\mathcal{P}(\xi^{\mathcal{O}}|\xi^{\mathcal{I}})\right)$. We use here the notations $\mathbb{E}[\cdot]$ and $\text{cov}(\cdot)$ to express respectively the expectation and covariance. The complete Gaussian Mixture Regression process is presented in Summary Box 2.2. Figure 2.6 provides a geometric illustration of the principle of this regression process. Equation (2.7) defines the conditional expectation of a multivariate normal distribution by estimating the conditional probability as a least squares estimate, i.e., the probability function $\mathcal{P}(\xi^{\mathcal{O}}|\xi^{\mathcal{I}})$ is represented by a single Gaussian distribution.

Gaussian Mixture Regression (GMR) can be used to retrieve smooth generalized trajectories with associated covariance matrices describing the variations and correlations across the different variables. In a generic regression problem, one is given a set of *predictor*, $\xi^{\mathcal{I}} \in \mathbb{R}^p$, and *response* variables, $\xi^{\mathcal{O}} \in \mathbb{R}^q$. The aim of regression is to approximate the conditional expectation of $\xi^{\mathcal{O}}$, given $\xi^{\mathcal{I}}$, on the basis of a set of observations $\{\xi^{\mathcal{I}}, \xi^{\mathcal{O}}\}$. For example, regression can be used to retrieve smooth trajectories that are generalized over a set of observed trajectories. The advantage of the approach is to provide a fast and analytic solution to produce smooth and reliable trajectories to be used for an efficient control of a robot (i.e., there is no need to manually verify or smooth the results).

For the particular example of trajectory learning, on the basis of a set of observations $\{\xi^{\mathcal{I}}, \xi^{\mathcal{O}}\}$, where $\xi^{\mathcal{O}}$ represents position vectors at time steps $\xi^{\mathcal{I}}$, the aim of the regression process is to estimate the conditional expectation of $\xi^{\mathcal{O}}$ for a given time step $\xi^{\mathcal{I}}$. This is followed by the computation of an expected

Summary Box 2.1 Properties of normal distributions.

Conditional Gaussian distribution:

Let $\xi \sim \mathcal{N}(\mu, \Sigma)$ be defined by

$$\xi = \begin{pmatrix} \xi^{\mathcal{I}} \\ \xi^{\mathcal{O}} \end{pmatrix}, \; \mu = \begin{pmatrix} \mu^{\mathcal{I}} \\ \mu^{\mathcal{O}} \end{pmatrix}, \; \Sigma = \begin{pmatrix} \Sigma^{\mathcal{I}} & \Sigma^{\mathcal{IO}} \\ \Sigma^{\mathcal{OI}} & \Sigma^{\mathcal{O}} \end{pmatrix}.$$

The conditional probability $\mathcal{P}(\xi^{\mathcal{O}}|\xi^{\mathcal{I}})$ is then defined by

$$\mathcal{P}(\xi^{\mathcal{O}}|\xi^{\mathcal{I}}) \sim \mathcal{N}(\hat{\mu}, \hat{\Sigma}), \tag{2.4}$$

$$\text{where} \quad \hat{\mu} = \mu^{\mathcal{O}} + \Sigma^{\mathcal{OI}}(\Sigma^{\mathcal{I}})^{-1}(\xi^{\mathcal{I}} - \mu^{\mathcal{I}}),$$

$$\hat{\Sigma} = \Sigma^{\mathcal{O}} - \Sigma^{\mathcal{OI}}(\Sigma^{\mathcal{I}})^{-1}\Sigma^{\mathcal{IO}}.$$

Linear transformation of Gaussian distribution:

If $\xi \sim \mathcal{N}(\tilde{\mu}, \tilde{\Sigma})$, the affine transformation $A\xi$ defined by matrix A follows the distribution

$$A\xi \; \sim \; \mathcal{N}(\mu, \Sigma), \tag{2.5}$$

$$\text{where} \quad \mu = A\tilde{\mu}, \qquad \Sigma = A\tilde{\Sigma}A^{\top}.$$

Product of Gaussian distributions:

The product of two Gaussian distributions $\mathcal{N}(\mu^{(1)}, \Sigma^{(1)})$ and $\mathcal{N}(\mu^{(2)}, \Sigma^{(2)})$ is defined by

$$C \cdot \mathcal{N}(\mu, \Sigma) = \mathcal{N}(\mu^{(1)}, \Sigma^{(1)}) \cdot \mathcal{N}(\mu^{(2)}, \Sigma^{(2)}), \tag{2.6}$$

$$\text{where} \quad C = \mathcal{N}(\mu^{(1)}; \mu^{(2)}, \Sigma^{(1)} + \Sigma^{(2)}),$$

$$\mu = \left((\Sigma^{(1)})^{-1} + (\Sigma^{(2)})^{-1}\right)^{-1}\left((\Sigma^{(1)})^{-1}\mu^{(1)} + (\Sigma^{(2)})^{-1}\mu^{(2)}\right),$$

$$\Sigma = \left((\Sigma^{(1)})^{-1} + (\Sigma^{(2)})^{-1}\right)^{-1}.$$

---

position, at each time step, to retrieve a generalized trajectory. GMR thus offers a means of extracting a single generalized trajectory comprised of a set of trajectories used to train the model. In these, the generalized trajectory is not part of the dataset but instead encapsulates all of its essential features. Figure 2.7 illustrates the regression process using the GMM representation.

Compared to traditional regression approaches, the regression function is not directly approximated. On the contrary, the joint density of the set of trajectories is first estimated by a model from which the regression function is derived. Modeling the joint density instead of a regression function has the advantage of providing a density distribution of the responses as opposed to of a single value. It thus simultaneously generates a mean response estimate, $\hat{\xi}$, and a covariance response estimate, $\hat{\Sigma}$, conditioned on the predictor variables, $\xi^{\mathcal{I}}$.

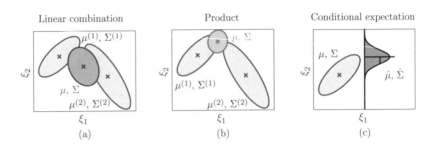

Figure 2.5 Properties of Gaussian density functions employed in the proposed model. (a) Linear combination $\mathcal{N}(\mu, \Sigma) \sim \mathcal{N}(\mu^{(1)}, \Sigma^{(1)}) + \mathcal{N}(\mu^{(2)}, \Sigma^{(2)})$. (b) Product $\mathcal{N}(\mu, \Sigma) \sim \mathcal{N}(\mu^{(1)}, \Sigma^{(1)}) \cdot \mathcal{N}(\mu^{(2)}, \Sigma^{(2)})$. (c) Conditional probability $\mathcal{P}(\xi_2|\xi_1) \sim \mathcal{N}(\hat{\mu}, \hat{\Sigma})$.

In order to learn trajectories, $\hat{\xi}$ is used as a generalized trajectory, while $\hat{\Sigma}$ is employed to extract the constraints on this trajectory. It thus permits the evaluation of the covariance information, not only at specific positions but also continuously along the movement.

Thus, the modeling phase becomes an important part of the algorithm in terms of processing, while the regression phase itself is processed very quickly. This is advantageous since the reproduction of smooth trajectories is fast enough to be used by the robot at any appropriate time, while the joint density modeling is computed in a phase of interaction that does not particularly need fast computation. Indeed, the estimation of $\mathbb{E}\left[\mathcal{P}(\xi^{\mathcal{O}}|\xi^{\mathcal{I}})\right]$ is simply calculated by a weighted sum of linear models.

Several encoding schemes can be considered as is further demonstrated in the various experiments presented in this book. For example, $\xi^{\mathcal{I}}$ may represent the motion of the right arm of a humanoid and $\xi^{\mathcal{O}}$ the motion of the left arm. The robot first learns how to carry out a bimanual task, requiring specific coordinations of the two arms and an encoding of the joint distribution $\mathcal{P}(\xi^{\mathcal{I}}, \xi^{\mathcal{O}})$ of the arm motion in a GMM. Then, if the user manipulates the right arm of the robot (e.g., to refine the learned skill), the robot can follow the motion appropriately with its left arm by estimating $\mathcal{P}(\xi^{\mathcal{O}}|\xi^{\mathcal{I}})$ through GMR. Similarly, the user can also, by estimating $\mathcal{P}(\xi^{\mathcal{I}}|\xi^{\mathcal{O}})$, manipulate the left arm while the robot simultaneously controls its right arm. The input/output variables can be changed on the fly as the model does not need to be retrained. Another approach is to use both spatial and temporal variables to explicitly describe a motion. This is done by considering a set of trajectories where each datapoint consists of a time value, t, and a position vector, x. Such an encoding scheme is used throughout this chapter to illustrate the regression process. It allows the retrieval of a smooth generalized version of the trajectories by encoding the joint distribution $\mathcal{P}(t, x)$ in a GMM, and then the retrieval of $\mathcal{P}(x|t)$ through GMR for a given range of time. An alternative strategy consists in directly encoding the dynamics of the motion, for example $\mathcal{P}(x, \dot{x})$ in the GMM, where x and $\dot{x}$ represent position and velocities. It thus becomes possible to obtain a motion by starting from a given position x and retrieving an estimation of the velocity with $\mathcal{P}(\dot{x}|x)$. As a consequence, the robot is brought to a new position.

Summary Box 2.2 The Gaussian Mixture Regression process.

A dataset of N datapoints of D dimensions is encoded in a *Gaussian Mixture Model* (GMM) of K Gaussians. The probability of a datapoint $\xi = [\xi^\mathcal{I}, \xi^\mathcal{O}]$ belonging to the GMM is defined by

$$P(\xi) = \sum_{k=1}^{K} \pi_k \, \mathcal{N}(\xi; \mu_k, \Sigma_k) = \sum_{k=1}^{K} \pi_k \, \frac{1}{\sqrt{(2\pi)^D |\Sigma_k|}} \, e^{-\frac{1}{2}\left((\xi-\mu_k)^\top \Sigma_k^{-1}(\xi-\mu_k)\right)},$$

where π_k are prior probabilities and $\mathcal{N}(\mu_k, \Sigma_k)$ represent Gaussian distributions defined by centers μ_k and covariance matrices Σ_k. Input and output components are represented separately as

$$\mu_k = \left[\begin{array}{c} \mu_k^\mathcal{I} \\ \mu_k^\mathcal{O} \end{array} \right], \quad \Sigma_k = \left[\begin{array}{cc} \Sigma_k^\mathcal{I} & \Sigma_k^\mathcal{IO} \\ \Sigma_k^\mathcal{OI} & \Sigma_k^\mathcal{O} \end{array} \right].$$

For a given input variable $\xi^\mathcal{I}$ and a given Gaussian distribution k, the expected distribution of $\xi^\mathcal{O}$ is defined by

$$P(\xi^\mathcal{O}|\xi^\mathcal{I}, k) \sim \mathcal{N}(\hat{\xi}_k, \hat{\Sigma}_k), \quad \text{where} \quad \begin{aligned} \hat{\xi}_k &= \mu_k^\mathcal{O} + \Sigma_k^\mathcal{OI}(\Sigma_k^\mathcal{I})^{-1}(\xi^\mathcal{I} - \mu_k^\mathcal{I}), \\ \hat{\Sigma}_k &= \Sigma_k^\mathcal{O} - \Sigma_k^\mathcal{OI}(\Sigma_k^\mathcal{I})^{-1}\Sigma_k^\mathcal{IO}. \end{aligned}$$

By considering the complete GMM, the expected distribution of $\xi^\mathcal{O}$, when $\xi^\mathcal{I}$ is known, can be estimated as

$$P(\xi^\mathcal{O}|\xi^\mathcal{I}) \sim \sum_{k=1}^{K} h_k \, \mathcal{N}(\hat{\xi}_k, \hat{\Sigma}_k),$$

where $h_k = P(k|\xi^\mathcal{I})$ is the probability that the Gaussian distribution k is responsible for $\xi^\mathcal{I}$

$$h_k = \frac{P(k)P(\xi^\mathcal{I}|k)}{\sum_{i=1}^{K} P(i)P(\xi^\mathcal{I}|i)} = \frac{\pi_k \, \mathcal{N}(\xi^\mathcal{I}; \mu_k^\mathcal{I}, \Sigma_k^\mathcal{I})}{\sum_{i=1}^{K} \pi_i \, \mathcal{N}(\xi^\mathcal{I}; \mu_i^\mathcal{I}, \Sigma_i^\mathcal{I})}.$$

By using the linear transformation property of Gaussian distributions defined in (2.5), the conditional expectation of $\xi^\mathcal{O}$, given $\xi^\mathcal{I}$, can be approximated by a single Gaussian distribution $\mathcal{N}(\hat{\xi}, \hat{\Sigma})$ with parameters

$$\hat{\xi} = \sum_{k=1}^{K} h_k \, \hat{\xi}_k, \quad \hat{\Sigma} = \sum_{k=1}^{K} h_k^2 \, \hat{\Sigma}_k. \tag{2.7}$$

By doing this recursively, the system can be used to reconstruct a generalized version of the motion without having to take time as an explicit parameter in the encoding scheme. This issue is discussed further in Chapter 5.

Although the theoretical considerations of GMR were presented in the machine learning literature several years ago (Ghahramani & Jordan, 1994;

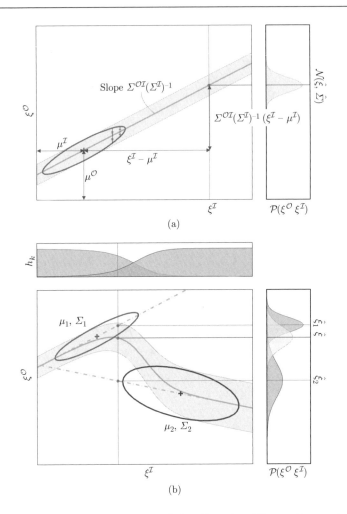

Figure 2.6 A graphical interpretation of the *Gaussian Mixture Regression* (GMR) process: (a) when considering a single Gaussian distribution; (b) when considering a GMM composed of two Gaussian distributions.

Cohn, Ghahramani, & Jordan, 1996; Sung, 2004; Kambhatla, 1996), the theory has come out with only few applications, which is surprising since GMM and associated EM learning algorithms are well established in various practical fields of research. Moreover, the approach is easy to implement, and satisfies robustness and smoothness criterions that are common to a variety of research fields, including robotics.

The number of Gaussians in the GMM determines the compromise for GMR between having an accurate estimation of the response and having a smooth response. This compromise is known as the *bias-variance tradeoff*. An appropriate model complexity must thus be determined in order for: (1) the bias between the estimate and the real values to be low (accuracy of the estimation with respect to the observed data); and (2) the variance on the estimate to also be low (smoothness of the estimation). Figure 2.8 illustrates the

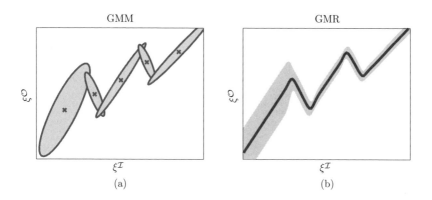

Figure 2.7 The *Gaussian Mixture Regression* (GMR) process used to compute the conditional density $\mathcal{P}(\xi^{\mathcal{O}}|\xi^{\mathcal{I}})$. The joint density, $\mathcal{P}(\xi^{\mathcal{I}}, \xi^{\mathcal{O}})$, is first modeled by a 2-dimensional GMM of 3 components (a). A sequence of input variables, $\xi^{\mathcal{I}}$, is then employed as query points to retrieve a sequence of output variables characterized by the center, $\hat{\xi}$, and covariances, $\hat{\Sigma}$, (b). The expected Gaussian distributions, $\mathcal{N}(\hat{\xi}, \hat{\Sigma})$, are thus represented along the trajectory as a generalized trajectory with an envelope representing the expected variations.

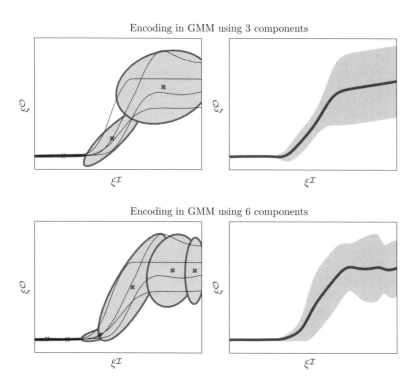

Figure 2.8 An illustration of the *bias-variance tradeoff* when defining the number of GMM components during encoding of the data (*left*) and retrieval of the data (*right*).

bias-variance tradeoff when defining the number of GMM components required for encoding and retrieving data. In this example, the model is more accurate with 6 Gaussians, whereas the generalized trajectory is smoother with 3 Gaussians.

A simpler approach to GMR would consist in directly computing the mean and standard deviation of the various demonstrations at each time step (see e.g. Ogawara et al. (2003)). This approach does however present several drawbacks: (1) the system is memory-based and needs to keep all historical data, which can lead to a scaling-up problem (see the rapid development of sensors for humanoid robots exploiting various modalities); (2) as PbD considers only a few demonstrations of the task, the use of simple statistics is usually insufficient for guaranteeing the generation of trajectories that are smooth enough to be reproduced by the robot; and (3) the constraints concerning the correlation across the different variables are not extracted.

2.5 REPRODUCTION BY CONSIDERING MULTIPLE CONSTRAINTS

In the preceding section, we discussed how to extract task constraints from a set of trajectories and how to reproduce a generalized version of these trajectories. We now consider the problem of finding an optimal controller for the robot when multiple independent constraints are considered and represented separately in various GMMs.

In this context (e.g., when considering actions on a variety of objects), the joint probability (i.e., the probability of two events in conjunction) is computed by estimating $\mathbb{E}\left[\mathcal{P}(\hat{\xi}^{(1)}) \cdot \mathcal{P}(\hat{\xi}^{(2)})\right]$ (independence assumption). By using standard statistical properties of Gaussian distributions, namely linear transformation, product and conditional distribution, we show in the following section that this estimation can be easily computed. We then see in Section 2.5.2 that it can also be interpreted as the minimization of a cost function measuring the similarity between the demonstrated and reproduced trajectories (optimization of a metric of imitation).

2.5.1 Direct computation method

After having extracted a set of M constraints represented by a series of Gaussian distributions $\{\hat{\xi}^{(h)}, \hat{\Sigma}^{(h)}\}_{h=1}^{M}$ through the GMR process, one needs to optimally combine the different constraints to reproduce the skill in a new situation. This is achieved by considering the product properties of normal distributions (see Summary Box 2.1 and Fig. 2.5).

The expected trajectory $\hat{\xi}$ is influenced by the set of generalized trajectories $\{\hat{\xi}^{(h)}\}_{h=1}^{M}$. However, the effect is not constant in time and the constraints are directly reflected by the covariance matrices $\{\hat{\Sigma}^{(h)}\}_{h=1}^{M}$. The expected trajectory, $\hat{\xi}$, and associated covariance, $\hat{\Sigma}$, can be computed by multiplying the various Gaussian distributions at each time step, i.e., by considering the constraints as independent and rewriting $\mathcal{P}(\xi) \sim \prod_{h=1}^{M} \mathcal{P}(\xi^{(h)})$ in terms of Gaussian distributions, namely

$$\mathcal{N}(\hat{\xi}, \hat{\Sigma}) \sim \prod_{h=1}^{M} \mathcal{N}(\hat{\xi}^{(h)}, \hat{\Sigma}^{(h)}).$$

When using the Gaussian product properties defined in (2.6), $\hat{\xi}$ and $\hat{\Sigma}$ can be computed with

$$\hat{\xi} = \left(\sum_{h=1}^{M} (\hat{\Sigma}^{(h)})^{-1} \right)^{-1} \left(\sum_{h=1}^{M} (\hat{\Sigma}^{(h)})^{-1} \hat{\xi}^{(h)} \right),$$

$$\hat{\Sigma} = \left(\sum_{h=1}^{M} (\hat{\Sigma}^{(h)})^{-1} \right)^{-1}. \tag{2.8}$$

An illustrative example for the reproduction process when considering multiple constraints is presented in Figure 2.9. The datasets $\xi^{(1)}$ and $\xi^{(2)}$ describing the two constraints are represented separately with two GMMs of 3 states (*first and second row*). GMR is used to retrieve the expected distributions typified by the Gaussian distributions $\mathcal{N}(\hat{\xi}^{(1)}, \hat{\Sigma}^{(1)})$ and $\mathcal{N}(\hat{\xi}^{(2)}, \hat{\Sigma}^{(2)})$ (depicted by a *dashed line* in the *third row*). The final expected distribution is then retrieved by multiplying the distributions of each constraint at each time step, i.e., by computing $\mathcal{P}(\xi) \sim \mathcal{P}(\xi^{(1)}) \cdot \mathcal{P}(\xi^{(2)})$, where $\mathcal{P}(\xi)$ corresponds to the Gaussian distribution $\mathcal{N}(\hat{\xi}, \hat{\Sigma})$. Through regression, a smooth trajectory, $\hat{\xi}$, satisfying the two sets of constraints is thus obtained (presented with a *solid line*) with associated constraints $\hat{\Sigma}$. Note that this process does not guarantee that the resulting trajectory is permissible (e.g., the presence of an obstacle during the reproduction).

2.5.2   Method based on optimization of a metric of imitation

In a similar manner, the result in (2.8) can be determined by deriving a metric of imitation. To measure the similarity between a candidate position ξ and a desired position $\hat{\xi}$, both of dimensionality D, a weighted cost function $\mathcal{H}(\xi) = (\xi - \hat{\xi})^{\top} W (\xi - \hat{\xi})$ is determined, where W is a $D \times D$ covariance matrix. By setting $W = \Sigma^{-1}$, the weight matrix is defined as the inverse of the covariance matrix observed during the demonstrations. In other words, if a variable is not consistent across the demonstrations (large variance), the associated weight is low, which indicates that this variable is not highly relevant for the reproduction of the task.

Let us consider a task where two constraints must be fulfilled, represented by two sets of Gaussian distributions $\{\hat{\xi}^{(1)}, \hat{\Sigma}^{(1)}\}$ and $\{\hat{\xi}^{(2)}, \hat{\Sigma}^{(2)}\}$. A metric of imitation Performance, $\mathcal{H}$, (i.e., the cost function for the task) is defined as

$$\mathcal{H}(\xi) = (\xi - \hat{\xi}^{(1)})^{\top} (\Sigma^{(1)})^{-1} (\xi - \hat{\xi}^{(1)})$$
$$+ (\xi - \hat{\xi}^{(2)})^{\top} (\Sigma^{(2)})^{-1} (\xi - \hat{\xi}^{(2)}). \tag{2.9}$$

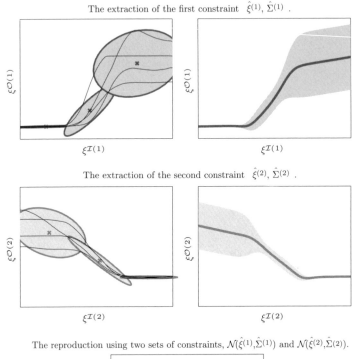

The extraction of the first constraint $\hat{\xi}^{(1)}$, $\hat{\Sigma}^{(1)}$.

The extraction of the second constraint $\hat{\xi}^{(2)}$, $\hat{\Sigma}^{(2)}$.

The reproduction using two sets of constraints, $\mathcal{N}(\hat{\xi}^{(1)},\hat{\Sigma}^{(1)})$ and $\mathcal{N}(\hat{\xi}^{(2)},\hat{\Sigma}^{(2)})$.

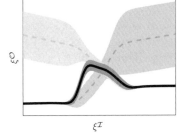

Figure 2.9 An illustrative example of the regression process when considering 2 independent constraints. *First and second row:* Independent extraction of two task constraints. *Third row:* Retrieval of a generalized trajectory and associated constraints (*solid line*) by combining information from the two independent constraints (*dashed line*).

To obtain an optimal trajectory, $\hat{\xi}$, minimizing $\mathcal{H}$, $\frac{\partial \mathcal{H}}{\partial \xi} = 0$ is solved. This is accomplished by using the second-order derivative properties of matrices (Petersen & Pedersen, 2008) and $W^\top = W$ (inverse covariance matrices are symmetric). If A is an affine transformation matrix and s is a vector, we have

$$\frac{\partial}{\partial x}(Ax - s)^\top W(Ax - s) = -2A^\top W(Ax - s),$$
$$\frac{\partial}{\partial x}Ax = A^\top.$$

These properties are used to derive the metric $\mathcal{H}$, which gives

$$-2\,(\Sigma^{(1)})^{-1}\,(\xi - \hat{\xi}^{(1)}) - 2\,(\Sigma^{(2)})^{-1}\,(\xi - \hat{\xi}^{(2)}) = 0$$

$$\Leftrightarrow \quad (\Sigma^{(1)})^{-1}\,\xi + (\Sigma^{(2)})^{-1}\,\xi = (\Sigma^{(1)})^{-1}\,\hat{\xi}^{(1)} + (\Sigma^{(2)})^{-1}\,\hat{\xi}^{(2)}$$

$$\Leftrightarrow \quad \xi = \left((\Sigma^{(1)})^{-1} + (\Sigma^{(2)})^{-1}\right)^{-1}\left((\Sigma^{(1)})^{-1}\,\hat{\xi}^{(1)} + (\Sigma^{(2)})^{-1}\,\hat{\xi}^{(2)}\right).$$

By extending this computation to M constraints, the solution is observed to resemble the one presented in Eq. (2.8).

2.6 LEARNING OF MODEL PARAMETERS

Up to now, the HMM and GMM parameters have been assumed to be known. The present section addresses the automatic estimation of these parameters. GMM and HMM estimations of the parameters in a batch mode are first described, i.e., when the entire dataset is employed to learn the models. Two update mechanisms are then proposed to incrementally learn the parameters in a GMM when the different demonstrations are provided one by one, i.e., by using only the latest observation to refine the model and by discarding this observation afterwards.

2.6.1 Batch learning of the GMM parameters

The theoretical analysis of GMMs and the development of related learning algorithms have been largely studied in machine learning (McLachlan & Peel, 2000; Vlassis & Likas, 2002; Dasgupta & Schulman, 2000). Training of the GMM parameters is usually performed in a batch mode (i.e., using the whole training set) through the *Expectation-Maximization* (EM) algorithm (Dempster & Rubin, 1977). This is a simple local search technique guaranteeing a monotone increase of the likelihood during optimization.

A dataset of N continuous datapoints, $\{\xi_j\}_{j=1}^{N}$, is considered, where each datapoint, ξ_j, is of dimensionality D. By taking into account Eq. (2.1), we define $p_{k,j}$ as the posterior probability $\mathcal{P}(k|\xi_j)$, computed using the *Bayes* theorem

$$\mathcal{P}(k|\xi_j) = \frac{\mathcal{P}(k)\mathcal{P}(\xi_j|k)}{\sum_{i=1}^{K}\mathcal{P}(i)\mathcal{P}(\xi_j|i)},$$

and E_k is defined as the sum of posterior probabilities (used to simplify the notation). The set of parameters $\{\pi_k, \mu_k, \Sigma_k, E_k\}$ describing the GMM are then estimated iteratively until convergence, starting from a rough

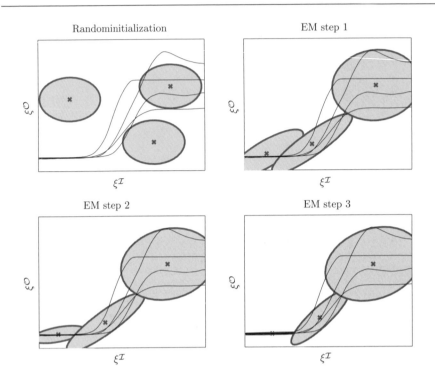

Figure 2.10 An illustration of the EM iterative learning algorithm. For this simple trajectory, the algorithm converges to a good local optimum already in 3 steps.

approximation of the parameters by *k-means* segmentation (MacQueen, 1967), as follows.

E-step:

$$p_{k,j}^u = \frac{\pi_k^u \, \mathcal{N}(\xi_j; \mu_k^u, \Sigma_k^u)}{\sum_{i=1}^K \pi_i^u \, \mathcal{N}(\xi_j; \mu_i^u, \Sigma_i^u)},$$

$$E_k^u = \sum_{j=1}^N p_{k,j}^u.$$

M-step:

$$\pi_k^{u+1} = \frac{E_k^u}{N},$$

$$\mu_k^{u+1} = \frac{\sum_{j=1}^N p_{k,j}^u \xi_j}{E_k^u},$$

$$\Sigma_k^{u+1} = \frac{\sum_{j=1}^N p_{k,j}^u (\xi_j - \mu_k^{u+1})(\xi_j - \mu_k^{u+1})^\top}{E_k^u}.$$

The iteration stops when the increase in the log-likelihood at each iteration becomes too small, i.e., when $\frac{\mathcal{L}^{u+1}}{\mathcal{L}^u} < C_1$. Here, the log-likelihood $\mathcal{L}$ is defined in Eq. (2.2). The threshold $C_1 = 0.01$ is set for further experiments.

Figure 2.10 illustrates the EM process employed for learning the GMM parameters. The initial centers of the Gaussian distributions (with constant diagonal covariance matrices) are voluntary set to inappropriate random values to
demonstrate the convergence performance of the algorithm.[3]

Other batch learning methods for GMM have been proposed as alternatives to the *Maximum Likelihood Estimation* (MLE). For example, Sung (2004)
set forth an *Iterative Pairwise Replacement Algorithm* (IPRA) for learning the
model parameters. This method iteratively merges similar pairs of components
and estimates the parameters of the new component with a method based
on moments, which is faster than EM update. This is of particular importance for the experiments presented by Sung (2004) since a very large number
of components is estimated. While the parameters found by IPRA represent
poor assessments in an MLE sense, the principal advantage of the approach
is its capability of evaluating the number of Gaussian K required to fit the
dataset without estimating MLE for multiple GMMs. In empirical studies, the
author showed that IPRA tended to produce smoother results when the mixture model was used for regression. However, he also showed that MLE training
of the model resulted in an improved classification performance. With regard
to classification, Sung (2004) thus suggested to use the IPRA approximation
of the parameters to initialize an MLE estimation.

As both recognition and reproduction issues are considered simultaneously in the applications proposed here, treating these issues in separate models
is not optimal and using MLE remains a coherent option for which the principal advantage consists in allowing the use of simple and flexible EM update
mechanism. Section 2.8.2 presents an alternative approach based on curvature
information to automatically select the number of components in a GMM prior
to its estimation through an EM algorithm. To produce smooth results when
the mixture model is used for regression, we also propose a method based on a
more simple regularization scheme, presented further in Section 2.9.1. In this
case, the sharper fit induced by the MLE estimation can be easily soften by
bounding the covariance matrices.

2.6.2   Batch learning of the HMM parameters

The HMM parameters are estimated through the *Baum-Welch* algorithm that
shares similarities with the EM technique presented in the previous section.
Even if the two learning processes (EM for GMM and EM for HMM) are
sometimes discussed separately, they can also be treated in the same framework
by using a similar notation (Bilmes, 1998; Xuan, Zhang, & Chai, 2001).

First, the *backward* variable, $\beta_{k,t}$, is defined as the probability of partially observing a sequence $\{\xi_t, \xi_{t+1}, \ldots, \xi_T\}$ of length $(T - t)$, given that the
current state is k at time t, i.e., $\beta_{k,t} = \mathcal{P}(\xi_t, \xi_{t+1}, \ldots, \xi_T | q_t = k)$. Similarly to
the *forward* variable previously defined in (2.3), $\beta_{k,t}$, is inductively computed
starting from $\beta_{k,T}$. By considering the parameters $\{\Pi, a, \mu, \Sigma\}$ of the HMM,

---

[3] Note that this convergence performance decreases when considering datasets of higher
dimensionality.

$\beta_{k,t}$ is calculated according to:

Initialization:

$$\beta_{k,T} = 1 \qquad \Big| \; \forall k \in \{1, \ldots, K\}.$$

Induction:

(2.10)

$$\beta_{k,t} = \sum_{i=1}^{K} a_{ki} \, \mathcal{N}(\xi_{t+1}; \mu_k, \Sigma_k) \, \beta_{i,t+1} \quad \left| \begin{array}{l} \forall k \in \{1, \ldots, K\} \\ \forall t \in \{T-1, T-2, \ldots, 1\}. \end{array} \right.$$

The variable $\gamma_{k,t}$ is then defined as the probability of being in state k at time t assuming that the observation sequence is ξ, i.e., $\gamma_{k,t} = \mathcal{P}(q_t = k|\xi)$. It is computed by considering the *forward-backward* variables defined in (2.3) and (2.10)

$$\gamma_{k,t} = \frac{\alpha_{k,t} \, \beta_{k,t}}{\mathcal{P}(\xi)} = \frac{\alpha_{k,t} \, \beta_{k,t}}{\sum_{i=1}^{K} \alpha_{i,t} \, \beta_{i,t}}.$$

Next, the variable $\zeta_{i,j,t}$ is defined as the probability of being in state i at time t and in state j at time $t+1$, given an observation sequence $\xi = \{\xi_1, \xi_2, \ldots, \xi_T\}$ of length T, i.e., $\zeta_{i,j,t} = \mathcal{P}(q_t = i, q_{t+1} = j|\xi)$. When the parameters $\{\Pi, a, \mu, \Sigma\}$ of the HMM are known, $\zeta_{i,j,t}$ is computed by considering the *forward-backward* procedure

$$\zeta_{i,j,t} = \frac{\alpha_{i,t} \, a_{ij} \, \mathcal{N}(\xi_{t+1}; \mu_j, \Sigma_j) \, \beta_{j,t+1}}{\mathcal{P}(\xi)}$$

$$= \frac{\alpha_{i,t} \, a_{ij} \, \mathcal{N}(\xi_{t+1}; \mu_j, \Sigma_j) \, \beta_{j,t+1}}{\sum_{i=1}^{K} \sum_{j=1}^{K} \alpha_{i,t} \, a_{ij} \, \mathcal{N}(\xi_{t+1}; \mu_j, \Sigma_j) \, \beta_{j,t+1}},$$

where the numerator represents $\mathcal{P}(q_t = i, q_{t+1} = j, \xi)$.

This summation of $\zeta_{i,j,t}$ over j is equivalent to the previously defined $\gamma_{k,t}$, i.e., $\gamma_{k,t} = \sum_{j=1}^{K} \zeta_{i,j,t}$. Summing $\zeta_{i,j,t}$ over t, i.e., $\sum_{t=1}^{T-1} \zeta_{i,j,t}$ gives rise to the expected number of transitions from state i to state j. Similarly, $\sum_{t=1}^{T-1} \gamma_{k,t}$ represents the expected number of transitions from state k.

If $\{\Pi^u, a^u, \mu^u, \Sigma^u\}$ are the current estimates of the HMM parameters, and by considering ζ^u and γ^u as additional variables computed from these evaluations, we can re-estimate the initial probability Π_k of being in state k, the states' transition probabilities a_{ij}, and the output distribution probabilities for each state by using

$$\Pi_k^{u+1} = \gamma_{k,1}^u \qquad \Big| \; \forall k \in \{1, \ldots, K\},$$

$$a_{ij}^{u+1} = \frac{\sum_{t=1}^{T-1} \zeta_{i,j,t}^u}{\sum_{t=1}^{T-1} \gamma_{i,t}^u} \qquad \left| \begin{array}{l} \forall i \in \{1, \ldots, K\} \\ \forall j \in \{1, \ldots, K\}, \end{array} \right.$$

$$\mu_k^{u+1} = \frac{\sum_{t=1}^{T} \gamma_{k,t}^u \xi_t}{\sum_{t=1}^{T} \gamma_{k,t}^u} \qquad \Big| \; \forall k \in \{1, \ldots, K\},$$

$$\Sigma_k^{u+1} = \frac{\sum_{t=1}^{T} \gamma_{k,t}^u (\xi_t - \mu_k^{u+1})(\xi_t - \mu_k^{u+1})^\top}{\sum_{t=1}^{T} \gamma_{k,t}^u} \Big| \; \forall k \in \{1, \ldots, K\}.$$

Linking with the EM estimation used for GMM (see Sect. 2.6.1), the variable $\gamma_{k,t}$ used for HMM corresponds to the posterior probability, $p_{k,j}$, in GMM.

2.6.3 Incremental learning of the GMM parameters

We have seen in Section 2.4 that a single demonstration of a skill is usually not sufficient to extract the task particularities and goals. Compared to a batch learning procedure, the benefits of an incremental learning approach is that the interaction can be performed in an online manner and that it becomes possible to observe the results immediately.

As seen in Section 2.6.1, the classic GMM learning procedure starts from an initial estimate of the parameters and uses the *Expectation-Maximization* (EM) algorithm to converge to an optimal local solution. This batch learning procedure is not optimal for PbD as it would be interesting to use new incoming data to refine the model of the gesture without having to keep the whole training data in memory.

The problem of incrementally updating the GMM parameters by taking into account only the new incoming data (and the previous estimation of the GMM parameters) has already been proposed for on-line data stream clustering. Song and Wang (2005) suggested the creation of a new Gaussian component to encode the new incoming data, as well as the fabrication of a compound model by merging the similar components of the previous GMM with the new GMM. The drawbacks of the suggested approach is the fact that it is computationally expensive and tends to produce more components than the standard EM algorithm. Arandjelović and Cipolla (2006) suggested the use of temporal coherence properties of data streams to update the GMM parameters. Their model assumes a smooth variation of the data with time, which allows an adjustment of the GMM parameters when new data are observed. The authors proposed a method to update the GMM parameters for each newly observed datapoint by allowing splitting and merging operations on the Gaussian components when the current number of Gaussians does not properly, model the new data.

In PbD, the issue is different in the sense that modeling on-line streams is not necessarily required. Nevertheless, an existing model needs to be refined when a new stream of datapoints has been observed and recognized by the model. Two approaches are suggested to update the model parameters: (1) a *direct update method* that takes inspiration from Arandjelović and Cipolla (2006) re-formulating the problem for a generic observation of multiple datapoints; and (2) a *generative method* that uses a batch learning process performed on data generated by *Gaussian Mixture Regression*. These two methods are described next.

Direct update method
The idea is to adapt the EM algorithm presented in Section 2.6.1 by separating the parts dedicated to the data which are already used to train the model with the one dedicated to the newly available data. The set of posterior probabilities $\{p(k|\xi_j)\}_{j=1}^{N}$ are assumed to remain the same when the new data $\{\tilde{\xi}_j\}_{j=1}^{\tilde{N}}$ are

used to update the model. It should be noted that this is true only if the new data are close to the model; a plausible assumption since the novel observation is first recognized by the model before being utilized for its re-estimation. In other words, the novel motion is similar to the ones already encoded in the model.

The model is initially created with N datapoints, ξ_j, and iteratively updated during T *EM-steps* (see Sect. 2.6.1), until convergence, to the set of parameters $\{\pi_k^U, \mu_k^U, \Sigma_k^U, E_k^U\}_{k=1}^K$. When a new motion is recognized by the model (see Sect. 2.2.1), $\tilde{U}$ *EM-steps* are again performed to adjust the model to the new $\tilde{N}$ datapoints, $\tilde{\xi}_j$, starting from the initial set of parameters $\{\tilde{\pi}_k^0, \tilde{\mu}_k^0, \tilde{\Sigma}_k^0, \tilde{E}_k^0\}_{k=1}^K = \{\pi_k^U, \mu_k^U, \Sigma_k^U, E_k^U\}_{k=1}^K$. The parameters $\{\tilde{E}_k^0\}_{k=1}^K$ are assumed to remain constant during the update procedure, i.e., it is supposed that the cumulated posterior probability does not change much with the inclusion of the novel data in the model. Using the notation $\tilde{p}_{k,j} = \mathcal{P}(k|\tilde{\xi}_j)$, the *EM* procedure can be reformulated as follows.

E-step:

$$\tilde{p}_{k,j}^u = \frac{\tilde{\pi}_k^u\, \mathcal{N}(\tilde{\xi}_j; \tilde{\mu}_k^u, \tilde{\Sigma}_k^u)}{\sum_{i=1}^K \tilde{\pi}_i^u\, \mathcal{N}(\tilde{\xi}_j; \tilde{\mu}_i^u, \tilde{\Sigma}_i^u)},$$

$$\tilde{E}_k^u = \sum_{j=1}^{\tilde{N}} \tilde{p}_{k,j}^u.$$

M-step:

$$\tilde{\pi}_k^{u+1} = \frac{\tilde{E}_k^0 + \tilde{E}_k^u}{N + \tilde{N}},$$

$$\tilde{\mu}_k^{u+1} = \frac{\tilde{E}_k^0 \tilde{\mu}_k^0 + \sum_{j=1}^{\tilde{N}} \tilde{p}_{k,j}^u \tilde{\xi}_j}{\tilde{E}_k^0 + \tilde{E}_k^u},$$

$$\tilde{\Sigma}_k^{u+1} = \frac{\tilde{E}_k^0 \left(\tilde{\Sigma}_k^0 + (\tilde{\mu}_k^0 - \tilde{\mu}_k^{u+1})(\tilde{\mu}_k^0 - \tilde{\mu}_k^{u+1})^\top \right)}{\tilde{E}_k^0 + \tilde{E}_k^u}$$

$$+ \frac{\sum_{j=1}^{\tilde{N}} \tilde{p}_{k,j}^u (\tilde{\xi}_j - \tilde{\mu}_k^{u+1})(\tilde{\xi}_j - \tilde{\mu}_k^{u+1})^\top}{\tilde{E}_k^0 + \tilde{E}_k^u}.$$

The iteration stops when $\frac{\mathcal{L}^{u+1}}{\mathcal{L}^u} < C_2$, where the threshold $C_2 = 0.01$ is set for further experiments.

Generative method

In the following, an alternative method to the direct update mechanism, which uses a stochastic approach to update the model parameters, is presented. An initial GMM with parameters $\{\pi_k, \mu_k, \Sigma_k\}_{k=1}^K$ is first created with the batch EM algorithm described in Section 2.6.1. Moreover, the corresponding GMR model is estimated as described in Section 2.4. When new data are available and recognized by the model, regression is performed on the current model to stochastically generate new data by considering the current GMR representation $\{\hat{\mu}_j, \hat{\Sigma}_j\}_{j=1}^T$. By creating a new dataset composed of these generated data and the new observed data $\{\tilde{\xi}_j\}_{j=1}^T$, the GMM parameters are refined again through the batch EM algorithm. For such a scenario, $\alpha \in [0; 1]$ is defined as the learning rate, and $n = n_1 + n_2$ is defined as the number of samples used for the learning procedure, in which $n_1 \in \mathbb{N}$ and $n_2 \in \mathbb{N}$ are respectively the number of examples duplicated from the new observation (copies of the observed trajectories) and the number of examples generated stochastically by the current model. The new training dataset is then defined by

$$\left. \begin{array}{ll} \xi_{i,j} = \tilde{\xi}_j & \text{if} \quad 1 < i \leq n_1 \\ \xi_{i,j} \sim \mathcal{N}(\hat{\mu}_j, \hat{\Sigma}_j) & \text{if} \quad n_1 < i \leq n \end{array} \right| \forall j \in \{1, \dots, T\},$$

where $n_1 = [n\alpha]$, and the notation $[\cdot]$ represents here the nearest integer function. The training set of n trajectories is subsequently employed to refine the model by updating the current set of parameters $\{\pi_k, \mu_k, \Sigma_k\}_{k=1}^{K}$ using the batch EM algorithm presented in Section 2.6.1. $\alpha \in [0, 1]$ can be set to a fixed learning rate or can depend on the current number of demonstrations used to train the model. In this case, when a new demonstration of $\tilde{N}$ datapoints is available (given that N datapoints from previous demonstrations were used to train the model), α is set to $\frac{\tilde{N}}{\tilde{N}+N}$. Identically, when each demonstration has the same number of datapoints, α can be recursively computed for each newly available demonstration. Starting from $\alpha^0 = 1$, the induction process can be defined as

$$\alpha^{u+1} = \frac{\alpha^u}{\alpha^u + 1}. \tag{2.11}$$

Figure 2.11 illustrates the stochastic incremental learning procedure, and Figure 2.12 depicts the encoding results of GMM during an incremental update

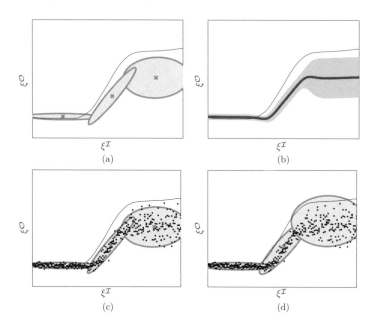

(a)

(b)

(c)

(d)

Figure 2.11 The incremental learning process when using the generative method. (a) The trajectory is recognized by the GMM. (b) GMR is performed on the current model. (c) The GMR model is used to stochastically retrieve a set of datapoints. (d) These datapoints and the recognized trajectory are employed to update the GMM parameters.

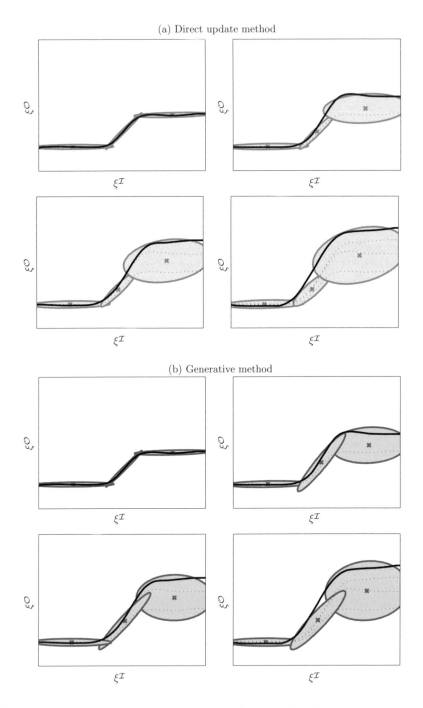

Figure 2.12 The incremental learning of data based on the *direct update method* (a) and the *generative method* (b).

of the parameters using respectively the *direct update method* and the *generative method*. The graphs show the consecutive encoding of the data in GMM, after convergence of the EM algorithms. Both algorithms use only the latest observed trajectory, represented with a solid line, to update the models (no historical data is employed). As can be seen, the two approaches converge to different local optima however the two solutions remain qualitatively very similar.

2.7 REDUCTION OF DIMENSIONALITY AND LATENT SPACE PROJECTION

In high dimensions (e.g., in the robot's joint space), the assessment of a GMM through EM becomes difficult due to the sparsity of the data when considering small numbers of datapoints. Indeed, as the likelihood function flattens, the difficulty of finding a local optimum that provides an efficient model of the data increases. This *curse of dimensionality* is a well-known problem of GMM (Scott, 2004).

To reduce the dimensionality of the data, we can benefit from the fact that the collected data may present redundancies. As the degree and type of redundancies can differ from one task to another, we seek a latent space onto which project the original dataset to find an optimal representation for the given task. For example, an optimal latent space for a writing task could be represented by a projection of the 3-dimensional original Cartesian position of the pen onto a 2-dimensional latent space defined by the writing surface. Similarly, a waving gesture could be represented as a simple 1-dimensional cyclic pattern.

In most applications, the optimal form of the functions to project human motion data is not known in advance. A latent space of motion extracted through a linear combination of the data in the original data space is assumed to be a sufficient pre-processing step for the further analysis of the movement. As mentioned in the introduction, *Principal Component Analysis* (PCA) is a common approach to project human motion data and has been used in various fields of research (Chalodhorn et al., 2006; Urtasun et al., 2004). This simple method can also be used in the proposed framework to reduce the dimensionality of the dataset for further analysis. We take the perspective that the whole dataset can first be projected globally into a latent space of motion, where the non-linearities of the signals are further represented by a local encoding of the data through GMM.

The linear decomposition of the data can be formulated as a *Blind Source Separation* (BSS) problem, where the observed trajectories x are assumed to be composed of statistically independent trajectories. The goal then consists in estimating the mixing matrix, A, to recover the underlying independent trajectories, ξ. To do so, we consider a linear transformation of the dataset x of N data points in the original centered *data space* of dimensionality d to a *latent space* of motion of dimensionality D, which produces a new dataset ξ of N datapoints. The transformation is defined by

$$x - \bar{x} = A\,\xi, \qquad\qquad (2.12)$$

where $\bar{x}$ represents the mean of the training dataset x and A is the transformation matrix. This goal cannot be perfectly achieved in practice due to a lack of general measures of statistical independence. However, other related criteria can be employed to approximate the decomposition into statistically independent components. Among the multiple criteria proposed to linearly decompose the data, we suggest using either:[4]

- *Principal Component Analysis* (PCA), which finds analytically a mixing matrix, A, projecting the dataset, x, onto uncorrelated components, ξ, with the criterion of ordering the dimensions with respect to the variability of the data.

- *Canonical Correlation Analysis* (CCA), which finds analytically a mixing matrix, A, projecting the dataset, x, onto a set of uncorrelated components, ξ, with the criterion of obtaining a maximum temporal correlation within each component.

- *Independent Component Analysis* (ICA), which finds iteratively a mixing matrix, A, projecting the dataset, x, onto independent components, ξ, with the criterion of having components that are as non-Gaussian as possible.

These methods make different assumptions on the dataset and can produce a variety of results for the projection. The assumptions depend on the type of signals and often represent the critical question of selecting one or the other method. PCA assumes that the important information is contained in the energy of the signals, CCA assumes that the important information is contained in the temporal ordering of the observations (and that the observations are composed of signals with varying autocorrelation functions), and ICA assumes that the components follow non-Gaussian distributions. The three methods are briefly described below.

2.7.1 *Principal Component Analysis* (PCA)

Principal Component Analysis (PCA) determines the directions along which the variability of the data is maximal (Jolliffe, 2002). This is done by the extraction of *eigenvectors* of the covariance matrix of the dataset x, $\Sigma^x = \mathbb{E}[xx^\top]$. The *eigenvectors*, v_i^x, and their associated *eigenvalues*, λ_i, are calculated by solving for

$$\Sigma^x v_i^x = \lambda_i v_i^x \qquad | \; \forall i \in \{1, \ldots, d\}.$$

The mixing matrix is then formed as $A = [v_1^x, v_2^x, \ldots, v_K^x]$, where K is the minimal number of *eigenvectors* used to obtain a satisfactory representation of

---

[4] For a survey on dimensionality reduction techniques, see, e.g., Fodor (2002).

the data, e.g., such that the projection of the data onto the reduced set of *eigenvectors* covers at least 98% of the data spread, $\sum_{i=1}^{K} \lambda_i > 0.98$. A is thus an orthogonal transformation that diagonalizes the covariance matrix, Σ^x.

2.7.2 Canonical Correlation Analysis (CCA)

Another means of separating mixed signals, x, is by *Canonical Correlation Analysis* (CCA) to find a linear combination of x that correlates most with a linear combination of the temporally delayed version of the signal $y = x(t+1)$ (Borga, 1998). By defining the linear combinations $\xi^x = (w^x)^\top x$ and $\xi^y = (w^y)^\top y$ ($\xi^x, \xi^y \in \mathbb{R}$), the correlations between ξ^x and ξ^y are given by[5]

$$\rho = \frac{\mathbb{E}[\xi^x \xi^y]}{\sqrt{\mathbb{E}[\xi^x (\xi^x)^\top] \, \mathbb{E}[\xi^y (\xi^y)^\top]}} = \frac{(w^x)^\top \, \Sigma^{xy} \, w^y}{\sqrt{((w^x)^\top \, \Sigma^x \, w^x) \, ((w^y)^\top \, \Sigma^y \, w^y)}},$$

where Σ^x and Σ^y are the covariance matrices within a set and Σ^{xy} is the covariance matrix between sets. Computing the maximum of ρ provides the directions of maximum data correlation, and setting the partial derivatives of ρ with respect to w^x and w^y to zero yields[6]

$$\left. \begin{array}{l} \left((\Sigma^x)^{-1} \, \Sigma^{xy} \, (\Sigma^y)^{-1} \, \Sigma^{yx}\right) \, w_i^x = \rho_i^2 \, w_i^x \\ \left((\Sigma^y)^{-1} \, \Sigma^{yx} \, (\Sigma^x)^{-1} \, \Sigma^{xy}\right) \, w_i^y = \rho_i^2 \, w_i^y \end{array} \right| \quad \forall i \in \{1, \ldots, d\}.$$

Hence, w_i^x and w_i^y are obtained by computing the *eigenvectors* of the matrices $\left((\Sigma^x)^{-1} \, \Sigma^{xy} \, (\Sigma^y)^{-1} \, \Sigma^{yx}\right)$ and $\left((\Sigma^y)^{-1} \, \Sigma^{yx} \, (\Sigma^x)^{-1} \, \Sigma^{xy}\right)$. The corresponding *eigenvalues*, ρ_i^2, are represented by the squared canonical correlation values. The *eigenvectors* corresponding to the largest *eigenvalue*, ρ_1^2, are the vectors w_1^x and w_1^y, which maximize the correlation between ξ^x and ξ^y. In other words, they maximize the autocorrelation (at lag one) of the resulting signal $\xi_1 = (w_1^x)^\top x$. Furthermore, w_2^x and w_2^y give the second best linear transformation providing the signal $\xi_2 = (w_2^x)^\top x$, which is uncorrelated with ξ_1. Following this, the linear transformation matrix, $A = [w_1^x, w_2^x, \ldots, w_K^x]$, can finally be constructed. A necessary condition for CCA includes varying autocorrelation functions for the source signals.

2.7.3 Independent Component Analysis (ICA)

Independent Component Analysis (ICA) defines a generative model where the data variables are assumed to be linear mixtures of unknown latent variables and where also the mixing system is unknown. The latent variables are assumed to be non-Gaussian and mutually independent, and are referred to as the independent components of the observed data. ICA is a recently developed variant of factor analysis that can perform *Blind Source Separation* (BSS), i.e., it

---

[5] By considering $y = x(t+1)$, a maximum autocorrelation approach is used. However, instead of taking into account only a data point and its next neighbor, CCA can be generalized to maximize the correlation between a data point and a linear combination of its neighboring region. In doing so, CCA not only finds an optimal combination of the signals but also obtains optimal filters for each signal.

[6] See Borga (1998) for the detailed computation.

is capable of iteratively extracting the independent components of the data. Thus, ICA finds a decomposition of the data minimizing a measure of Gaussianity (e.g. kurtosis, negentropy, mutual information measures), which can be interpreted as producing components, ξ, that are as independent as possible. In many practical situations, it also has the effect of reducing the statistical dependence of the data.

ICA assumes that the components follow a non-Gaussian distribution. This assumption can be illustrated by considering the addition of two Gaussian distributions. Indeed, as seen previously, such an addition is another Gaussian distribution. Therefore, if signals with Gaussian distributions are combined, it becomes difficult to extract the original signals from the resulting one. Similarly, we have seen that the linear projection of a multi-dimensional Gaussian distribution produces another Gaussian distribution. These two facts suggest that a way of identifying the "interesting" features in a multi-dimensional dataset consists in retrieving the projections onto which the signals are as non-Gaussian as possible. While the aim of PCA is to find projections onto which the signals present high variance (second order statistics), ICA gives rise to projections onto which the signals present high kurtosis (fourth order statistics). A fixed-point iteration scheme, also referred to as the *FastICA* algorithm, can be used to compute an estimation of ICA (Hyvärinen, 1999). This iteration scheme is one of the most widely used algorithms for performing independent component analysis due to its computational efficiency in combination with its statistical robustness.

ICA requires pre-whitening of the data, which is done in a pre-processing step by PCA. During this pre-processing stage, the dimensionality is also reduced by discarding the components with the lowest variance. The mixing matrix, A, is square and is iteratively updated from a usually randomly selected starting point. The components ξ found by ICA are presented in a random order. Depending on the application, the re-ordering of the components with respect to their measure of Gaussianity does not always provide satisfying results. To prevent this, the mixing matrix, A, found by PCA is used as an initial estimate, with the principal advantage of providing the same starting point at each run. As PCA already finds a latent space of motion through second order statistics, it can speed up ICA convergence as a result of a large portion of the job already having been performed by PCA. In other words, ICA can concentrate on higher-order moments. The drawback is that, by initializing the iterative search with an estimate composed of second-order statistics, the process may also in some cases converge to a poor local optimum.

2.7.4   Discussion on the different projection techniques

Figure 2.13 illustrates the use of PCA, CCA and ICA to find an optimal latent space representation of the motion and to reduce the dimensionality of the dataset. In this example, we see that CCA and ICA could present an advantage for the further encoding of the data in GMM or HMM. Indeed, the various segments characterizing the trajectories are better separated when using CCA and ICA, which may be advantageous for further processing.

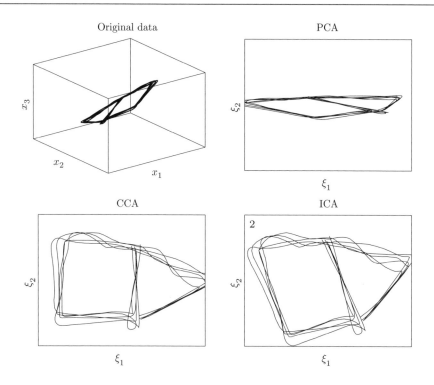

Figure 2.13 An illustration of a 3-dimensional dataset projected onto a 2-dimensional latent space by either PCA, CCA and ICA.

PCA, CCA and ICA share the common constraint of requiring a linear transformation A producing mutually uncorrelated components ξ. Uncorrelatedness is a reasonable constraint since the components, ξ, are assumed to be independent, but it is still a weaker assumption than statistical independence. Uncorrelatedness can be achieved by a variety of linear transformations, and therefore requires another criterion for a solution to be obtained. While PCA maximizes variance (the energy of the signal), ICA minimizes Gaussianity or other related measure, and CCA maximizes autocorrelation of the resulting components.

One disadvantage of PCA involves it being strongly dependent on the range of values (or magnitude) of the various components. This is a severe drawback when considering a robot endowed with multiple sensors collecting information and where the variance of each sensory component is unrelated to the importance of the received information. For example, if one would collect Cartesian and joint angle data and perform PCA on this dataset, the directions of the principal components could vary depending on the joint angle measure adopted (e.g., degrees or radians). Even by using a single modality (e.g., joint angles), PCA can lose important information concerning the joints with small ranges of motion since each joint angle is characterized by its own range of values. For example, the shoulder humeral rotation (axial rotation) usually demonstrates low variability but remains very important for manipulation skills (e.g., to pass an object from one hand to the other). A solution to this issue would be to normalize the various joint angles with respect to their magnitude. However, such a measure can be task-dependent or difficult to properly define. It should also be noted that, with PCA, the first component provides information regarding the direction of the following component (perpendicular to the first direction), which is not the case for ICA, for which the components are truly independent, i.e., no information concerning the second component is given by the first.

Correlation measures the relations across components that are invariant to their magnitude. While both PCA and ICA take into account the shapes of the distribution of the data (variance and non-Gaussianity), CCA focuses on the temporal characteristic of the signals. Indeed, an important property of canonical correlations is their invariance to affine transformation, and, thus, to scalings of the signals. In both PCA and ICA, the temporal correlations are not taken into account, i.e., the solutions remain unchanged if the data are arbitrarily re-ordered. Thus, depending on the data, ICA makes the BSS problem unnecessarily difficult. By ignoring the temporal relations within the signals, relevant information is discarded. This is a drawback when analyzing human motion data, since such signals have important temporal information causing their autocorrelation. CCA could in such cases be a good candidate for solving the BSS problem with less computational effort by utilizing this autocorrelation information.

In spite of the theoretical considerations, it should here be noted that the experiments conducted with real-world data, presented in Section 3.2, reveal that CCA and ICA have in fact very few advantages over PCA when considering simple gestures (at least when the data are further processed by GMM or HMM).

2.8   MODEL SELECTION AND INITIALIZATION

So far, we have assumed that the optimal number of Gaussians in the GMM (or HMM) was known *a priori*. In this section, we discuss how this can be evaluated, and address the subsequent problem of initializing the Gaussian distributions so that *Expectation-Maximization* (Sect. 2.6.1) can converge to an appropriate local optimum. We first present one usual criterion for model selection, *Bayesian Information Criterion* (BIC), and then suggest an alternative approach based on curvature information in order to take advantage of the continuous properties of human gestures.

2.8.1   Estimating the number of Gaussians based on the *Bayesian Information Criterion* (BIC)

A drawback of EM is that the optimal number of components K in a model may not be known beforehand. A tradeoff must thus be determined between optimizing the model likelihood (a measure of how well the model fits the data) and minimizing the number of parameters (i.e., the number of states used to encode the data). One common method for doing so consists in estimating multiple models with an increasing number of components, and selecting an optimum based on certain model selection criteria (Vlassis & Likas, 2002). For this purpose cross validation is a common method, however it has the disadvantage of requiring additional demonstrations to form a test set, which is inconvenient in PbD. To avoid this process, several criteria based on information theory have been proposed; among them, the *Bayesian Information Criterion* (BIC) is one of the most commonly used (Schwarz, 1978).[7] The BIC score is defined as

$$S_{\text{BIC}} = -\mathcal{L} + \frac{n_p}{2} \log(N),$$

where $\mathcal{L}$ is the log-likelihood of the model (see Sects. 2.2.1 and 2.3.1) using the demonstrations as a testing set, n_p is the number of free parameters required for a GMM of K components, i.e., $n_p = (K - 1) + K \left(D + \frac{1}{2}D(D + 1)\right)$ for a GMM with full covariance matrices, and N is the number of D-dimensional datapoints. The first term of the equation measures how well the model fits the data, while the second term is a penalty factor aiming at minimizing the number of parameters. Thus, multiple GMMs are estimated and the BIC score is used to select the optimal number of Gaussian distributions K.

2.8.2   Estimating the number of Gaussians based on trajectory curvature segmentation

The main drawback of BIC is that it requires the estimation of multiple models. Sung (2004) showed that the optimal number of parameters, K, for regression is usually lower than for density estimation, which can be illustrated by observing that regression when employing different values of K can result in the same regression function. In this section, we tackle the model selection paradigm in a similar spirit by using the specific particularities of temporally continuous trajectories, and analyze the curvature of these trajectories to provide information on the number of components required for further encoding in GMM.

As noted by Ghahramani and Jordan (1994), when encoding trajectories, the mixture of Gaussian distributions competitively partitions the input space and learns a linear regression surface in each partition. Each Gaussian distribution encodes a portion of the trajectory, which is quasi-linear in a hyperspace, i.e., the first *eigenvalue* of the covariance matrix is very large compared to the remaining values (usually more than one order of magnitude). The associated *eigenvector* is mainly directed toward the temporal axis (a point that is further discussed in Sect. 2.9.2). Consequently, the portions of trajectories in between two Gaussian distributions generally present sharper turns.[8] By considering

---

[7] Note that this criterion is similar to that of the *Minimum Description Length* (MDL).

[8] Note that this statement is true for most of the human movements considered here, but that it nonetheless cannot deal with a particular subset of signals in 2D and 3D with constant generalized curvatures (very regular movements such as circles or helixes).

temporally continuous trajectories, we thus developed an alternative to the k-*means* initialization of GMM for the particular case of continuous trajectories, by using an initialization process based on the segmentation of the signals with generalized curvature information. Advantages was then taken of the temporal information encapsulated in the representation, i.e., by using the sequential ordering of the points and the continuous characteristics of the motion data.

Let us consider a D-dimensional curve $\xi(t)$, for which a *Frenet frame*, which is a moving frame used to naturally describe local properties of the curve in terms of a local reference system, is defined. This is done by the set of orthonormal vectors $\{e_1(t), e_2(t), \ldots, e_D(t)\}$, constructed using the *Gram-Schmidt* orthogonalization process. The process starts with one basis vector $e_1(t)$, defined by the first derivative of $\xi(t)$

$$q_1(t) = \frac{\partial \xi(t)}{\partial t},$$
$$e_1(t) = \frac{q_1(t)}{||q_1(t)||}. \tag{2.13}$$

The following basis vectors are computed by orthogonally projecting the j-th derivative of $\xi(t)$ using the previous basis vectors in a recursive manner. With (2.13), the rest of the vectors $\{e_j(t)\}_{j=2}^{D}$ can then be recursively computed with

$$q_j(t) = \frac{\partial^j \xi(t)}{\partial t^j} - \sum_{i=1}^{j-1} e_i(t)^\top \left(\frac{\partial^j \xi(t)}{\partial t^j} \right) e_i(t),$$
$$e_j(t) = \frac{q_j(t)}{||q_j(t)||}. \tag{2.14}$$

Using the parameters defined in (2.13) and (2.14), the generalized curvatures $\{\chi_j(t)\}_{j=1}^{(D-1)}$ are defined by

$$\chi_j(t) = \frac{e_{j+1}(t)^\top \left(\frac{\partial e_j(t)}{\partial t} \right)}{||\frac{\partial \xi(t)}{\partial t}||}. \tag{2.15}$$

For a D-dimensional curve $\xi(t)$, this process involves the computation of the D-th derivatives of $\xi(t)$, and the first derivative of $\{e_j(t)\}_{j=1}^{(D-1)}$. For this, the curve is locally approximated by a D-dimensional polynomial function, using a Gaussian window of variance $\frac{T}{100}$. The derivatives can subsequently be analytically computed and are re-sampled to provide trajectories of T datapoints.[9]

For the special case of 3-dimensional paths, $\{e_1(t), e_2(t), e_3(t)\}$ define respectively the *tangent*, *normal* and *binormal* vectors. Moreover, $e_3(t)$ can be

---

[9] It should be noted that since $\xi(t)$ is represented as a finite sampling of 100 points on average, the approximation of these derivatives by numerical differentiation would not provide satisfactory estimations.

computed using $e_3(t) = e_1(t) \times e_2(t)$. The first generalized curvature $\chi_1(t) = \kappa(t)$ is called *curvature* and measures the deviance of $\xi(t)$ from a straight line relative to the osculating plane. The second generalized curvature $\chi_2(t) = \tau(t)$ is called *torsion* and measures the deviance of $\xi(t)$ from a plane curve. If the *torsion* is null, the curve lies completely in the osculating plane. *Curvature* is thus a measure of the rotation of the *Frenet frame* about the binormal vector $e_3(t)$, whereas *torsion* is a measure of the rotation of the *Frenet frame* about the tangent vector $e_1(t)$. The *Darboux vector* $\omega(t)$ provides a concise way of interpreting *curvature* and *torsion* geometrically, and is defined by $\omega(t) = \tau(t)e_1(t) + \kappa(t)e_3(t)$ with the norm $\|\omega(t)\| = \sqrt{\tau(t)^2 + \kappa(t)^2}$ representing the *total curvature* (Kehtarnavaz & deFigueiredo, 1988). Similarly, with Eq. (2.15), we define the norm $\Omega(t)$ for a generic D-dimensional curve as

$$\Omega(t) = \sqrt{\sum_{i=1}^{D-2} \chi_i(t)^2}.$$

This value captures the multiple features of the curve $\xi(t)$ in terms of generalized curvatures (Kehtarnavaz & deFigueiredo, 1988). Thus, the local maxima of $\Omega(t)$ provide points for segmenting the trajectory,[10] where the data between two segmentation points represent parts of the trajectory for which directions do not vary much. The initial means and covariances $\{\mu_k, \Sigma_k\}_{k=1}^{K}$ are then computed by considering the various portions of the signals separated by two consecutive segmentation points. The number of Gaussians used in the GMM is thereafter directly defined by the number of segmentation points minus one. Figure 2.14 presents an example of the segmentation process based on the total curvature information.

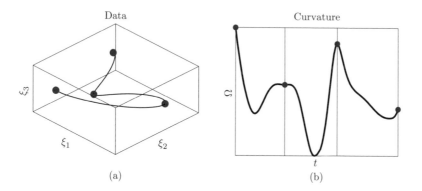

(a) (b)

Figure 2.14 An illustration of the curvature-based segmentation process used to initialize the GMM parameters. (a) Segmented data in 3D Cartesian space where each segmentation point defines a sharp turn in the motion. (b) The total curvature trajectory used to segment the data by extracting local maxima (represented by points).

---

[10] The beginning and the end of the trajectory are also considered as segmentation points.

In a batch-learning mode, when considering multiple demonstrations, the number of segmentation points can vary, as can their positions. In this case, *k-means* clustering can be used to generalize over the different examples, where the number of clusters is defined by the most common number of segmentation points extracted for the different trajectories. A multivariate Gaussian distribution is then fit to each portion of the trajectories between two centers of the clusters.

2.9   REGULARIZATION OF GMM PARAMETERS

In this section, three simple but very helpful regularization schemes are introduced and will be considered throughout the experiments when estimating the GMM parameters.

2.9.1   Bounding covariance matrices during estimation of GMM

In a GMM, when the variables are locally highly dependent, the covariance matrix encoding this portion of the signal is close to a singular matrix, resulting in unstable parameter estimates. The associated drawbacks are twofold: (1) the trajectory retrieved by regression can present sharp edges and bumpy transitions between two neighboring Gaussians; (2) EM can fail to update the GMM correctly when new data are available (the covariance matrices close to singularities are stuck to their current estimates). To ensure a numerical stability of the covariance matrix, the simple regularization scheme suggested here consists in adding a constant diagonal matrix at each iteration of the EM algorithm. This signifies that, in the update mechanism described in Section 2.6.1, Σ is replaced by $\Sigma + \delta I$, where I is the identity matrix and δ is a scalar proportionally defined to the variance of the dataset. Figure 2.15 illustrates the advantage of bounding the covariance matrices for the regression process.

2.9.2   Single mode restriction during reproduction through GMR

When encoding in GMM trajectories with explicit time variables, we suggest an adequate stretching of the values in order for the temporal variables to have a larger range of values than their spatial counterparts. This can be set by simply considering different units for the representation of the dataset. For example, regarding joint angles, $t \in \{1, 2, \ldots, T\}$ is defined with $T = 100$ as the temporal range, while the joint angles are expressed in radians, i.e., with a maximum range of values $\theta \in [-\pi, \pi]$. The aim of this rescaling process is to get rid of the multi-modal distributions that may appear during the GMR process. An output of multi-modal distributions is not a modeling error *per se*, but the multimodal distributions are not appropriately tackled by GMR. Indeed, at each time step, a single Gaussian distribution is estimated to

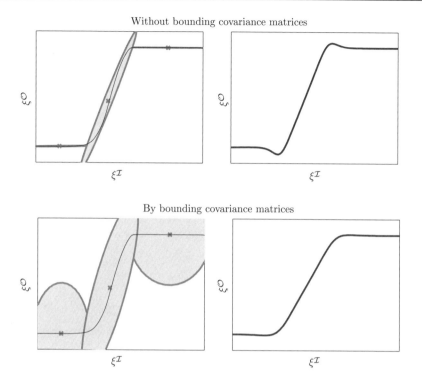

Figure 2.15 An illustration of the regression problem that may be encountered when one or multiple covariance matrices in the model are close to singular. *First row:* Training with a classic EM algorithm and reproduction through GMR. *Second row:* Training with an EM algorithm, involving bounding the covariance matrices at each iteration, and reproduction through GMR. We see in the first row that the transitions between neighboring Gaussians overfit the corners around time steps 50 and 80, and that this effect is attenuated when bounding the covariance matrices (second row).

model the multimodal distribution. As a single covariance estimation is used to approximate the distribution of data, it may in these situations become difficult to efficiently extract the task constraints.

Figure 2.16 demonstrates that initializing the GMM parameters with *k-means* (using a random initialization of the cluster centers) is influenced by the range of values used to describe the temporal and spatial components. The example presented in this figure shows a system failing at efficiently encoding the data if the variability of the input components (i.e. time t) is inappropriate. In the first row, due to the range of values for the temporal component, t, being smaller than for the spatial component, EM converges to a solution that is correct in a likelihood sense but that presents poor regression properties. Indeed, when performing GMR, multiple Gaussian distributions with similar weights describe the output variable ξ at each time step, t. The expected distributions at these time steps present multimodal distributions (multiple peaks), which

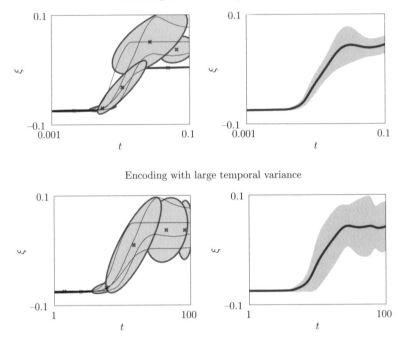

Figure 2.16 An illustration of the multimodal distribution issue (multi-peaked distributions) when estimating the GMM parameters with EM using *k-means* with random centers at initialization. We see that the resulting GMM depends to a large extent on the unit used for the temporal input values (resulting in a smaller or larger time variance).

are estimated by a single distribution through GMR. The results thus display a relatively correct generalized trajectory, but the associated covariance matrices fail to capture relevant information on the task constraints. In the second row, as the range of values for the input component, t, is one order of magnitude larger than for the output component, ξ, the expected distribution when performing GMR at each time step is close to being unimodal, i.e., the expected distribution presents a single peak. Note that the number of Gaussians have deliberately been overestimated to show the influence of the temporal variance on the nature of the distributions retrieved by regression, and that the burden resulting from multi-peaked distributions is often attenuated when only a low number of Gaussian components is considered.

Another solution to avoid this issue would be to change the *k-means* algorithms by distributing the initial cluster centers equally in time. Indeed, initialization by *k-means* is not always optimal when considering temporally continuous trajectories due to the centers of the clusters initially being defined at random. Given that trajectories bounded in time are considered, initial estimations of the centers μ can be obtained by distributing the initial clusters equally in time. This is achieved by taking into account each subset of

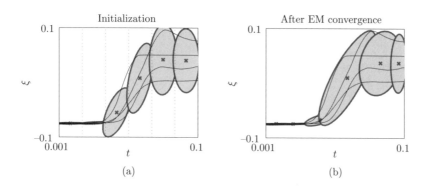

(a)   (b)

Figure 2.17 An illustration of the importance of initializing the GMM parameters to avoid the problem of multi-peaked distributions when performing GMR. Here, the GMM parameters are initialized using regular temporal intervals to prevent the multi-peaked effect when retrieving data through GMR. We see that the estimation of the parameters is robust to the different ranges of temporal values. As a comparison, we refer to Figure 2.16 where the GMM is initialized by k-means with random centers.

data satisfying $\frac{(i-1)T}{K} < t \leq \frac{iT}{K}$, and by using them to compute μ_i and Σ_i analytically for each Gaussian distribution $i \in \{1, \ldots, K\}$ (see Fig. 2.17(a)). Figure 2.17(b) shows that the multi-peaked distributions resulting from the EM learning process disappear when initializing the GMM parameters with regular temporal intervals. Thus, the trajectories resulting from GMR become insensitive to the temporal scaling factor.

2.9.3 Temporal alignment of trajectories through *Dynamic Time Warping* (DTW)

It was demonstrated in Section 2.3 that *Hidden Markov Models* (HMMs) could be used to encapsulate the temporal variations of the trajectories previously encoded in *Gaussian Mixture Models* (GMMs). Section 2.4 suggested the use of GMM/GMR as an appropriate approach to extract the constraints in a continuous form by encoding trajectories in a mixture of Gaussians. For the particular case of trajectories defined by time components and spatial components, GMM treats the temporal and spatial variables indifferently, which can be used to analytically retrieve a smooth trajectory through regression.

It is advantageous to utilize GMR over the HMM stochastic reproduction process since GMR provides a fast and analytic means of reconstructing smooth trajectories. This advantage is discussed in detail in Section 3.1. However, GMR still has the disadvantage of not being very robust to the temporal variability across the different demonstrated trajectories. Indeed, the use of GMR is mainly relevant for extracting variability and correlation information concerning the spatial values, but could, ideally, be useful in some cases to automatically align the demonstrations prior to further processing.

By way of HMM, this could be automatically carried out by modeling the temporal variability through transition probabilities. To provide a
similar effect with regard to the GMM/GMR approach, the trajectories presenting strong non-homogenous temporal distortions can still be temporally
re-aligned with pattern-based approaches. When required, *Dynamic Time
Warping* (DTW) can thus be used as a template matching pre-processing step
to temporally align the trajectories (Chiu, Chao, Wu, & Yang, 2004). DTW
is sometimes considered as a weaker method as opposed to HMM, but it has
the advantage of being simple and robust, and can be used conjointly with
GMR. DTW gives rise to a non-linear alignment of the demonstrated trajectories with respect to a reference trajectory. A distance array is first computed
and a DTW path is searched by a dynamic programming approach through
the array, usually by considering slope limits to prevent degenerate warps. To
improve the computational efficiency of the process, the alignment can be performed in a latent space of motion of lower dimensionality than the original data
space.

Let us consider two trajectories, ξ^A and ξ^B, of length T. We define the
distance between two datapoints of temporal index k_1 and k_2 by $h(k_1, k_2) =
\|\xi^A_{k_1} - \xi^B_{k_2}\|$. A warping path, $S = \{s_l\}_{l=1}^L$, defined by L elements $s_l = \{k_1, k_2\}$,
is subject to several constraints. *Boundary conditions* are given by $s_1 = \{1, 1\}$
and $s_L = \{T, T\}$. If $s_l = \{a, b\}$ and $s_{l-1} = \{a', b'\}$, *monotonicity* is provided
by $a \geq a'$ and $b \geq b'$, while *continuity* is defined by $a - a' \leq 1$ and $b - b' \leq 1$.
Dynamic programming is employed in order to minimize $\sum_{l=1}^{L} s_l$, by expressing
the cumulative distance $\gamma(k_1, k_2)$ as the distance $h(k_1, k_2)$ found in the current
cell and the minimum of the cumulative distances of the adjacent elements by
defining the following recursion.

Initialization:

$$\gamma(1, 1) = 0.$$

Induction:

$$\gamma(k_1, k_2) = h(k_1, k_2) + \min \left[\gamma(k_1 - 1, k_2 - 1), \gamma(k_1 - 1, k_2), \gamma(k_1, k_2 - 1) \right].$$

Global and local constraints are usually provided to reduce the computational cost of the algorithm and to limit the permissible warping paths.
Here, we experimentally fixed an *adjustment window condition* and a *slope
constraint condition* defining the maximum amount of warping allowed, as in
Sakoe and Chiba (1978). For a known warping path, it is possible to align one
of the trajectories by taking the other as reference. In the proposed framework, GMR is used to retrieve a trajectory from the current GMM model,
which is subsequently employed as a reference trajectory to align those newly
observed.

Figure 2.18 shows an example of temporal alignment of various demonstrations where the model is trained incrementally. For each newly observed
trajectory, a generalized trajectory is computed by regression based on the
current model and is then used as a reference trajectory to re-align the one
observed. The GMM is updated with this aligned observation. Without

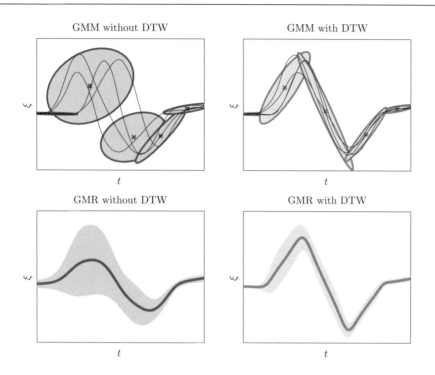

Figure 2.18 An estimation of GMM parameters (*top*) and the GMR retrieval process (*bottom*) with raw data (*left*) and after pre-processing by *Dynamic Time Warping* (*right*).

DTW pre-processing (*left*), the system has difficulties in extracting the task constraints due to the strong distortion in time of the trajectories. The generalized trajectory retrieved through GMR still follows the main characteristics of the demonstrated trajectories but the local minimum and maximum of the original data are not attained, i.e., the peaks of the retrieved trajectory are much too softened. With DTW (*right*), the essential features of the trajectories are maintained, and the generalized trajectory retrieved through GMR keeps the local minimum and maximum as observed during the demonstrations.

2.10   USE OF PRIOR INFORMATION TO SPEED UP THE LEARNING PROCESS

We have seen in Sections 2.4 and 2.5 that it is possible to extract the constraints of a task through GMR and that multiple constraints can be considered by multiplying the resulting Gaussian distributions or by deriving a metric of imitation performance to find a controller that simultaneously takes into consideration both constraints. We now discuss how *a priori* information on the task constraints can be incorporated in the system to alternate the relative importance of these constraints without having to use only statistics. Equation (2.9) in Section 2.5.2 is then reformulated to add *a priori* information on the constraints

$$\mathcal{H}(\xi) = w_1 \, (\xi - \hat{\xi}^{(1)})^\top \, (\Sigma^{(1)})^{-1} \, (\xi - \hat{\xi}^{(1)})$$
$$+ w_2 \, (\xi - \hat{\xi}^{(2)})^\top \, (\Sigma^{(2)})^{-1} \, (\xi - \hat{\xi}^{(2)}). \qquad (2.16)$$

Here, w_1 and w_2 define respectively the weights of the first and second constraints considered for the task. Similarly, by using the direct computation method presented in Section 2.5.1, prior information can be defined for each constraint by directly multiplying each covariance matrix by the inverse of each weight.

Figure 2.19 presents the use of priors to modify the reproduction of a skill composed of two separate constraints (see also Fig. 2.9). The direct computation method with covariances $w_1^{-1}\Sigma^{(1)}$ and $w_2^{-1}\Sigma^{(2)}$ is used to find a controller for the reproduction of the task. One should note that the weights satisfy the relation $w_1^{-1}w_2^{-1} = 1$.

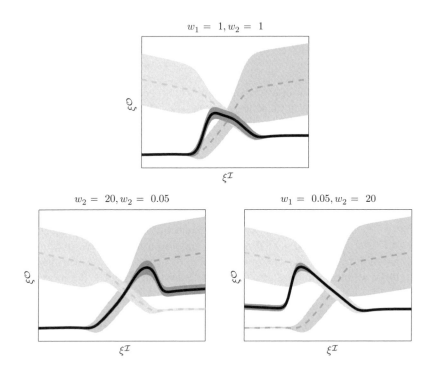

Figure 2.19 An illustration of the retrieval process through GMR when using different manually-set priors. The first graph displays the reproduction results from the example presented in Figure 2.9, where the constraints are extracted through statistics only. To manually modify the influences of the various constraints, w_1 and w_2 can then be set to different values (last two graphs).

One possible application is to fix priors varying along the motion in order to smoothly blend various constraints. Another application further described in Chapter 7 is to combine information collected through various forms of social cues with the statistical information extracted through cross-situational observations.

2.11 EXTENSION TO MIXTURE MODELS OF VARYING DENSITY DISTRIBUTIONS

GMM is a powerful tool for encoding continuous signals. However, as the sensory system of the robot is not solely composed of this type of signals, it is important for a generic probabilistic framework to have the possibility of dealing with signals of various forms. Indeed, mixture modeling is a popular approach for density approximation of continuous as well as of binary or discrete data, and Gaussian distribution is but a special case of density functions in a *mixture of experts* framework (McLachlan & Peel, 2000). We here present an extension of the model to the use of binary signals. This is of particular interest for the robots considered in further experiments since the sensors and actuators of the end-effectors are represented by such binary signals (on/off commands), e.g., where the two-dimensional binary trajectories correspond to the status of the left and right hand of a humanoid robot.

2.11.1 Generalization of binary signals through a *Bernoulli Mixture Model* (BMM)

A dataset of N binary datapoints, $\{\xi_j\}_{j=1}^N$, is considered, where each datapoint, ξ_j, is of dimensionality D, i.e., $\xi_j = \{\xi_{j,i}\}_{i=1}^D$. To encode this dataset in a probabilistic framework, a *mixture of multivariate Bernoulli distributions* can be used similarly to a *mixture of multivariate Gaussian distributions* for encoding continuous data (Carreira-Perpiñán, 2002). The dependencies across the binary data are captured thanks to the contribution of the components of the mixture.

Similarly to Eq. (2.1), a mixture of D-dimensional Bernoulli density functions of parameters or *prototypes* $\{\mu_{k,i}\}_{i=1}^D$, with $\mu_{k,i} \in [0,1]$, is defined by $\mathcal{P}(\xi_j) = \sum_{k=1}^K \mathcal{P}(k)\mathcal{P}(\xi_j|k)$ with

$$\mathcal{P}(k) = \pi_k,$$
$$\mathcal{P}(\xi_j|k) = \mathcal{B}(\xi_j; \mu_k)$$
$$= \prod_{i=1}^D (\mu_{k,i})^{\xi_{j,i}} (1 - \mu_{k,i})^{1-\xi_{j,i}}.$$

Analogously to the approximation of GMM parameters (see Sect. 2.6.1), $\{\pi_k, \mu_k\}$ are estimated by first defining the variables $p_{k,j} = \mathcal{P}(k|\xi_j)$ and

$E_k^u = \sum_{j=1}^{N} p_{k,j}^u$, and then employing the EM algorithm to learn the parameters iteratively.

E-step:

$$p_{k,j}^u = \frac{\pi_k^u \, \mathcal{B}(\xi_j; \mu_k^u)}{\sum_{i=1}^{K} \pi_i^u \, \mathcal{B}(\xi_j; \mu_i^u)},$$

$$E_k^u = \sum_{j=1}^{N} p_{k,j}^u.$$

M-step:

$$\pi_k^{u+1} = \frac{E_k^u}{N},$$

$$\mu_k^{u+1} = \frac{\sum_{j=1}^{N} p_{k,j}^u \xi_j}{E_k^u}.$$

Similarly to Gaussian distribution, each Bernoulli distribution $\mathcal{B}(\mu_k)$ can be defined by its moments, represented by the center, μ_k, and covariance matrix, $\Sigma_k = \mathrm{diag}(\mu_k(1 - \mu_k))$ (see Carreira-Perpiñán (2002) for details). The Gaussian Mixture Regression (GMR) process presented in Section 2.4 can thus correspondingly similarly be applied to BMM. Figure 2.20 presents an

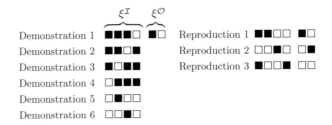

Figure 2.20 An example of binary signals encoding in *Bernoulli Mixture Model* (BMM) and retrieval through *Bernoulli Mixture Regression* (BMR). A robot collects 6-dimensional binary data when learning how to grasp an object. The first two values correspond to the activation of two bumpers within the robot's right hand. The next two values represent bumpers within the robot's left hand. The fourth and fifth values make up the status of the robot's right and left hand (opening or closing). In this example, the robot must learn that the two bumpers need to be activated simultaneously to close the hand, which corresponds to the state when the object to grasp is centered in the left or right palm. The 6-dimensional binary data is encoded in a BMM of 3 Bernoulli distributions (which intuitively may correspond to the activation of the right or left hand and to a neutral state where neither hand is activated). When presenting again a sequence of 4-dimensional binary inputs $\xi^\mathcal{I}$, the system retrieves an estimation of $\xi^\mathcal{O}$ that is used to open/close the right and left hand. If the same patterns as those used for training are presented, the system correctly retrieves the output data. Moreover, if novel signals that are not part of the training set are presented, we see that the system is still able to correctly generalize the skill.

illustrative example of encoding through BMM and retrieval through *Bernoulli Mixture Regression* (BMR).

2.12 SUMMARY OF THE CHAPTER

This chapter presented various methods for encoding multiple demonstrations of a motion in a probabilistic framework. It was shown that a probabilistic representation of the skill can be used to continuously extract the relevant constraints, and to reproduce the learned skill by generalizing over the demonstrations encapsulated in the model.

Section 2.1 presented an illustration of the proposed probabilistic approach, and Section 2.2 discussed the use of a *Gaussian Mixture Model* (GMM) and *Hidden Markov Model* (HMM) to encode a set of continuous movements. Section 2.4 described the reproduction issue through *Gaussian Mixture Regression* (GMR), and Section 2.5 demonstrated how to reproduce trajectories when multiple constraints are considered simultaneously. Section 2.6 addressed the learning issue, where Expectation-Maximization was first presented and where two incremental learning strategies were then proposed to learn the GMM parameters without keeping historical data in memory. Section 2.7 introduced several techniques for reducing the dimensionality of the data by projecting the trajectories in a latent space of motion. Section 2.8 discussed the selection of the optimal number of Gaussians in the GMM by way of the *Bayesian Information Criterion* (BIC) or by using a curvature-based segmentation process. Section 2.9 presented various means of regularizing the GMM parameters by bounding the covariance matrices to retrieve smoother trajectories and by employing *Dynamic Time Warping* (DTW) as a pre-processing step to deal with the temporal distortions across multiple demonstrations. Section 2.10 discussed the use of prior information in the probabilistic framework, and Section 2.11 finally addressed the possible extensions of the model to mixture models of varying density distributions such as the *Bernoulli Mixture Model* (BMM).

The datasets utilized in this chapter consisted of simple low-dimensional trajectories to illustrate the processing steps. The applications of these algorithms to human-robot interaction scenarios with real-world human motion data are presented in the following chapters.

CHAPTER 3

COMPARISON AND OPTIMIZATION OF THE PARAMETERS

This chapter presents preliminary experiments concerned with the testing and optimization of the several algorithms proposed theoretically in Chapter 2, in which human motion data is utilized to validate their use in a human-robot interaction framework.

Experiments on reproduction through HMM and GMM/GMR, latent space projection, model selection and incremental learning strategies are presented, and these methods are principally aimed at stimulating the use of the various algorithms in the human-robot teaching experiments that will be presented in subsequent chapters. The reader who would prefer to move directly to the essential applications of the PbD framework can jump forward to the experiments presented in Chapter 4.

3.1 OPTIMAL REPRODUCTION OF TRAJECTORIES THROUGH HMM AND GMM/GMR

This experiment concerns the problem of retrieving a smooth trajectory from an HMM by using a variety of approaches and compares the results with the explicit encoding of temporal and spatial components in a GMM as well as the retrieval through GMR.

3.1.1 Experimental setup

3 demonstrations of human motion in Cartesian space are provided in a simple test set. A series of motion sensors recorded the user's gestures by collecting joint angle trajectories of the upper-body torso. The movements were recorded by 8 *X-Sens* motion sensors attached to the torso, upper-arms, lower-arms, hands (at the level of the fingers) and back of the head.[1] Each sensor provided

---

[1] These motion sensors are commercially available, see http://www.xsens.com for more information.

the 3D absolute orientation of each segment by integrating the 3D rate-of-turn, acceleration and earth-magnetic field at a rate of 50 Hz and with a precision of 1.5 degrees. The data were sent to a computer for further processing by wireless *Bluetooth* communication (or alternatively through a *USB* connection). For each joint, a rotation matrix was defined as the orientation of a distal limb segment expressed in the frame of reference of its proximal limb segment. The kinematics motion of each joint could then be computed by decomposing the rotation matrix into joint angles (see Figs. 3.1 and 3.2).

The motion sensors returned orientation matrices $R_{W\rightarrow 1}$ and $R_{W\rightarrow 2}$ representing respectively the orientation of the proximal and distal segments, both expressed in a static frame of reference, W. The orientation of the distal

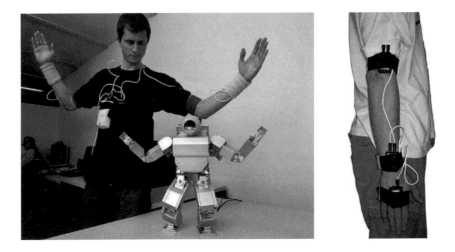

Figure 3.1 An illustration of the use of *X-Sens* motion sensors to record the user's gesture by collecting joint angle trajectories of the upper-body torso.

Figure 3.2 Three successive rotations describing the motion of the shoulder joint following a Cardanic order convention (XYZ decomposition order). This decomposition has the advantage of being interpreted in terms of anatomical motion defined by a sequence of 3 angles, representing *flexion-extension*, *abduction-adduction* and *humeral rotation* of the shoulder.

segment with respect to its proximal counterpart was given by $R_{1\to2}$ through the relation

$$R_{W\to2} = R_{W\to1}R_{1\to2}$$
$$\Leftrightarrow R_{1\to2} = (R_{W\to1})^{-1}R_{W\to2}.$$

Calibration of the motion sensors
For each joint, a desired calibration posture was defined by a rotation matrix R_0, where the set of matrices formed a natural posture of the body. During the calibration process, the user was requested to set his/her upper body in the same posture, R_0, by copying the position of a stick figure displayed on a screen. To be easily reproducible, the posture R_0 corresponded to a standing position where the person was looking forward, with the arms hanging along the body. The rotation matrices recorded by the sensors during the calibration phase were defined by $\tilde{R}_0$. The difference between the desired calibration posture, R_0, and the recorded calibration posture, $\tilde{R}_0$, was obtained by expressing $\tilde{R}_0$ in the frame of reference defined by R_0, i.e., by computing $R_\Delta = R_0\tilde{R}_0^{-1}$. After calibration, for each new posture acquired, we defined $R_\Delta = R\tilde{R}^{-1}$, where $\tilde{R}$ is the current rotation matrix recorded by the motion sensors. For further processing, we employed the set of rotation matrices, R, taking into account the rotation error R_Δ

$$R = R_0\tilde{R}_0^{-1}\tilde{R}.$$

Determining joint angle trajectories from rotation matrices
Each rotation matrix, R, was decomposed by a succession of 3 rotations along various axes. A decomposition order following the *Cardanic convention* (also known as the *Tait-Bryant convention*) was considered, and was defined by 3 consecutive rotations, $R_x(\phi) \to R_y(\theta) \to R_z(\psi)$, around axes X, Y' and Z''

$$R = R_z(\psi)R_y(\theta)R_x(\phi)$$

$$= \begin{pmatrix} c_\psi & -s_\psi & 0 \\ s_\psi & c_\psi & 0 \\ 0 & 0 & 1 \end{pmatrix} \begin{pmatrix} c_\theta & 0 & s_\theta \\ 0 & 1 & 0 \\ -s_\theta & 0 & c_\theta \end{pmatrix} \begin{pmatrix} 1 & 0 & 0 \\ 0 & c_\phi & -s_\phi \\ 0 & s_\phi & c_\phi \end{pmatrix}$$

$$= \begin{pmatrix} c_\psi c_\theta & -s_\psi c_\phi + c_\psi s_\theta s_\phi & s_\psi s_\phi + c_\psi s_\theta c_\phi \\ s_\psi c_\theta & c_\psi c_\phi + s_\psi s_\theta s_\phi & -c_\psi s_\phi + s_\psi s_\theta c_\phi \\ -s_\theta & c_\theta s_\phi & c_\theta c_\phi \end{pmatrix}.$$

Here, the rotation angles ϕ, θ and ψ represented respectively the *roll*, *pitch* and *yaw* angles ($s.$ and $c.$ correspond to $\sin(\cdot)$ and $\cos(\cdot)$ functions, respectively). By decomposing the rotation matrix into

$$R = \begin{pmatrix} r_{11} & r_{12} & r_{13} \\ r_{21} & r_{22} & r_{23} \\ r_{31} & r_{32} & r_{33} \end{pmatrix},$$

we then find

$$\theta = \sin^{-1}(-r_{31}), \qquad \phi = \tan^{-1}\left(\frac{r_{32}}{r_{33}}\right), \qquad \psi = \tan^{-1}\left(\frac{r_{21}}{r_{11}}\right), \qquad (3.1)$$

with an alternative set of joint angles given by[2]

$$\theta = \pi - \sin^{-1}(-r_{31}), \qquad \phi = \tan^{-1}(\frac{r_{32}}{r_{33}}) + \pi, \qquad \psi = \tan^{-1}(\frac{r_{21}}{r_{11}}) + \pi.$$

Figure 3.2 presents an illustration of this process applied to the right shoulder joint. The human arm is here defined as a kinematic chain where the *glenohumeral joint* (or *shoulder joint*) connects the girdle and the upper arm (3 DOFs), the *elbow* connects the upper arm and the forearm (1 DOF), the *wrist* connects the forearm and the hand (1 DOF) and the fingers are connected to the hand (1 DOF). However, it is important to keep in mind that joints in a human arm are not ideal and do not hold a reference location to the segments they connect. In most robotics systems, one usually assumes for the sake of simplicity that centers of rotation exist despite the fact that no joint in a human arm meets this criterion (Gams & Lenarcic, 2006).

Each *X-Sens* motion sensor provided an orientation matrix expressed in a fixed frame of reference, which is a complete description of the orientation without *gimbal locks*.[3] However, due to the redundancy of the rotation matrix representation, the orientation was defined by 9 parameters and it was difficult to interpret and process the data efficiently (there existed no physical interpretation for the parameters in the rotation matrix and it was not possible to directly handle velocity). To analyze human motion data, the biomechanical approach consisted of defining a *Joint Coordinate System* (JCS) and an associated angle convention permitting the analysis of joint movements in terms of anatomical rotations. Joint angles are defined as the orientation of one segment coordinate system relative to another. Thus, by providing a hierarchy of the various joints, the joint angles can be processed in a similar way for each body part, as presented above.

The rotation matrix can be constructed by applying 3 consecutive rotations around specified axes. By considering several sequences of rotations, various triplets of angles can be extracted. The selected sequence of successive rotations depends mainly on the analysis to be performed. Typically there exist two types of successive rotations: Eulerian and Cardanian. The Eulerian type involves a repetition of rotations about one particular axis: XYX, XZX, YXY, YZY, ZXZ, ZYZ, whereas the Cardanian type is characterized by the rotations about all three axes: XYZ, XZY, YZX, YXZ, ZXY, ZYX. In this experiment, the XYZ decomposition order was selected. Even though the Cardanian type differed from the Eulerian type in terms of the combination of rotations, they both use a very similar approach to compute the orientation angles. As described above, the *Cardanic convention* is defined by 3 consecutive rotations $R_x(\phi) \rightarrow R_y(\theta) \rightarrow R_z(\psi)$ around axis X, Y' and Z'', where the rotation angles $\{\phi, \theta\, \psi\}$ define respectively the *roll*, *pitch* and *yaw*. In contrast, the *Eulerian convention* is defined by 3 consecutive rotations

---

[2] Using a trace of the previous joint angles, the most appropriate set of joint angles could be selected at each time step by keeping the values corresponding the most to the ones computed at the previous time step.

[3] A configuration defined by Euler angles where one loses one dimension of rotation.

$R_z(\phi) \rightarrow R_x(\theta) \rightarrow R_z(\psi)$ around axes Z, X' and Z'', where the rotation angles are given as $\{\psi, \theta\,\phi\}$, representing the *precession*, *nutation* and *spin* angles. Unfortunately, for joints such as shoulders, there is no single definition of the joint angles that are anatomically meaningful for the full range of motion that the joints can achieve. There is thus no standard sequence of rotations for describing the shoulder motion, and each of the representations suffers from *gimbal locks*.[4]

The convention order can be selected to place the singularity beyond the motion ranges that are typically recorded during the experiment. To do so, the *International Shoulder Group* (ISG) has published recommendations aimed at standardizing the reporting of kinematic data to allow a facilitated comparison of results among researchers (Wu et al., 2005). This group recommends that the shoulder angle (upper arm relative to torso) should preferably be described by the Cardanic representation including flexion-extension, abduction-adduction and humeral (or axial) rotation used in this experiment. However, due to the range of motion of the shoulder being considerable, they also recommend to select the rotation sequence for the task under analysis as there exists no sequence of rotations that is anatomically meaningful throughout the workspace of the shoulder. Here, the principle reason for selecting the Cardanic order convention was to match the XYZ joint hierarchy of the robot used in further experiments.

In the present experiment, the set of joint angles was utilized to reconstruct the position of the hand in 3D Cartesian space through direct kinematics. To do so, the lengths of several body parts were estimated to create a kinematic model of the user's body.

3.1.2 Experimental results

A body of work has suggested to employ a *stochastic* approach to retrieve human motion sequences from HMM (Inamura, Kojo, & Inaba, 2006; Inamura et al., 2005; D. Lee & Nakamura, 2006). In this method, the initial states distribution and transition probabilities of the HMM are first used to stochastically generate multiple sequences of states. For each of these, the corresponding output distributions defined by the Gaussian distributions then give rise to the stochastic generation of multiple trajectories. Generalized trajectories can finally be retrieved by averaging over a large amount of generated sequences. Figure 3.3 presents reproduction results obtained through this averaging approach.

Another technique consists in using the HMM to encode keypoints from multiple continuous trajectories previously extracted in a pre-processing phase (Calinon & Billard, 2004; Calinon et al., 2005; Asfour et al., 2006; Aleotti & Caselli, 2006b, 2005). This way, a gesture is considered as a sequence of relevant events in the trajectory. For example, inflexion points can be extracted either by considering zero-velocity crossing or by employing a more elaborated

---

[4] This can be seen by observing the components of the matrix in Eq. (3.1), expressed with cosine and sine, which leads to expectations of finding non-unique solutions.

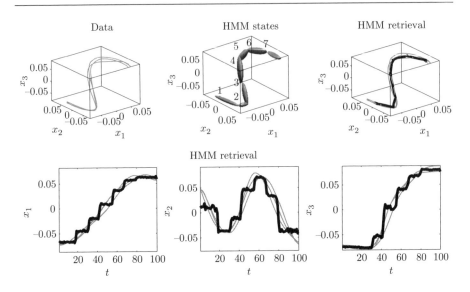

Figure 3.3 The encoding of raw data using HMM and the retrieval based on an averaging approach (for 100 generated trajectories). The retrieved trajectories were not very smooth due to the stochastic nature of the process.

segmentation scheme based on curvature information (see Sect. 2.8.2). This segmentation process is in adequation with several studies highlighting the presence of typical correlation patterns between joint angles at distinct phases of a manipulation process (Popovic & Popovic, 2001). For reproduction, the most likely sequence of states is first retrieved by the model and a third order spline fit interpolation is used to obtain a continuous generalized trajectory. The use of a *piecewise cubic Hermite spline*[5] gives rise to a smooth trajectory that is guaranteed to lie within the robot's range of motion (Fritsch & Carlson, 1980). Also referred to as *clamped cubic spline*, this interpolation function ensures BIBO stability of the system (see Calinon et al. (2005) for details).[6] Thus, under bounded disturbances, the original signal remains bounded and does not diverge, i.e., the trajectory resides within the range defined by the keypoints. By an appropriate extraction of the keypoints, it becomes possible to retrieve trajectories using only a few HMM states.

Figure 3.4 presents the reproduction results for the keypoint approach. For one of the demonstrations, a keypoint has been detected near the middle of the trajectory (corresponding to state 3), due to a slight perturbation having

---

[5] A *cubic Hermite spline* is a third-degree spline where each polynomial of the spline is defined in Hermite form, consisting of two control points and two control tangents for each polynomial. A *piecewise cubic Hermite spline* guarantees monotonicity in each interval defined by two keypoints, while maintaining smooth interpolation properties.

[6] BIBO stands for *Bounded-Input Bounded-Output*, and is generally used in signal processing and control theory as a stability condition. In other words, if a system is stable, the output will be bounded for every input to the system that is bounded (W. Sun, 1999).

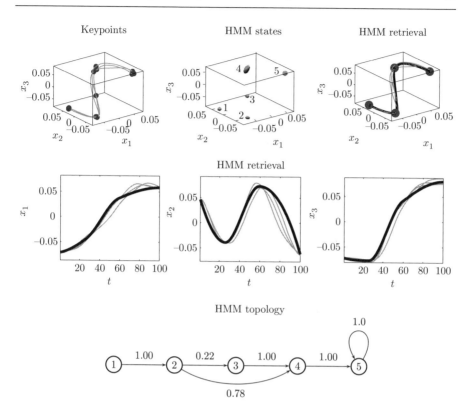

Figure 3.4 The encoding of a sequence of keypoints using an HMM of 5 states and the retrieval through interpolation (represented with a thick black line). The original dataset is given with thin grey line. The last row displays the topology learned by the model described by transition probabilities (see also Table 3.1).

occurred during this demonstration. The overall process was however unperturbed by this single outlier. The transition probabilities and initial state distributions of the HMM are reported in Table 3.1. As the keypoints reflect important changes in the trajectories, modeling of these trajectories in HMM by employing these keypoints and retrieving a generalized version of the trajectories through interpolation usually requires fewer states as opposed to encoding of the raw data. The transition probabilities show that the non-zero values for each state are mostly distributed to a single value (transition to the next keypoint), except for state 2, since state 3 has not always been detected as a keypoint during demonstrations. In other words, the transition from state 2 to state 4 is the one that is most likely to be observed. The most probable sequence of states $\{1, 2, 4, 5\}$ is then used for the retrieval process.

A similar approach, also based on interpolation but that directly encodes the continuous data, has been proposed (Calinon & Billard, 2007c, 2005; Billard et al., 2006). In this method, the reproduction of a gesture is triggered by the user, i.e., the gesture is reproduced after recognition of a similar one performed

Table 3.1 Transitions probabilities and initial states distribution parameters of the HMM encoding keypoints presented in Figure 3.4.

Transition probability a_{ij}					
i \ j	1	2	3	4	5
1	0.00	**1.00**	0.00	0.00	0.00
2	0.00	0.00	**0.22**	**0.78**	0.00
3	0.00	0.00	0.00	**1.00**	0.00
4	0.00	0.00	0.00	0.00	**1.00**
5	0.00	0.00	0.00	0.00	**1.00**

Initial states probability Π_i				
1	2	3	4	5
1.00	0.00	0.00	0.00	0.00

by the demonstrator. An optimal sequence of states is first retrieved from the recognized model by way of the *Viterbi* algorithm presented in Section 2.3.1 (alternatively, one or multiple sequences of states could also be stochastically retrieved). A sequence of states of size T is thus extracted, and is subsequently reduced to a subset of N states with an associated duration representing the number of steps where we remain in the same state. From this sequence of N states, a set of N keypoints is retrieved, represented by a sequence of centers, μ. By knowing the duration in-between two keypoints, a generalized version of the trajectory can finally be reconstructed with the *piecewise cubic Hermite polynomial* interpolation.

Figure 3.5 illustrates such a reproduction process, and Figures 3.6 and 3.7 present the reproduction results for this continuous approach, where HMMs with 5 and 7 states are considered. The parameters of the first HMM are also reported in Table 3.2.

Finally, Figures 3.8 and 3.9 present the reproduction results for the GMM/GMR approach where 5 and 7 Gaussians are considered. In comparison to the other approaches, the generalized trajectories extracted through regression can be seen to, qualitatively, fit the data very well, which is confirmed by the quantitative analysis presented in Figure 3.10. From this graph, the GMR approach is found to be much less sensitive to the number of states as opposed to the HMM continuous approach.

Table 3.3 presents a summary of the advantages/disadvantages of the various approaches based on theoretical and computational points of view. The *HMM stochastic approach* can be computationally expensive and the smoothness of the resulting trajectory cannot be guaranteed due to the stochastic nature of the retrieval process. Another side-effect is a tendency for the local minima and maxima of the signals to be cut-off and smoothed (which are often essential features to reproduce). In order to be viable, this reproduction process usually requires a higher number of states in the HMM than for recognition purposes.

Retrieval of the most likely sequence of states

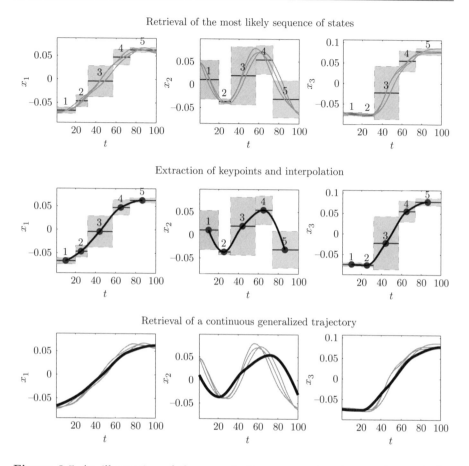

Extraction of keypoints and interpolation

Retrieval of a continuous generalized trajectory

Figure 3.5 An illustration of the reproduction process based on an interpolation technique and by using HMM to encode raw data. *First row:* A dataset used to train the HMM (*grey solid line*) as well as the retrieval of the most likely sequence of states, where the output distributions are represented as boxes depicting centers and variances. *Second row:* The extraction of keypoints from the sequence of states (*points*), and spline fit interpolation (*black solid line*). *Third row:* The retrieval of a generalized version of the data by resizing the interpolated trajectory in time.

The *HMM keypoint approach* also has several drawbacks: (1) it heavily relies on an efficient segmentation of the trajectories which can be application-dependent and requires fine-tuning of the segmentation parameters; (2) only the centers, μ, of the distribution are used to generate the generalized keypoints, signifying that the correlation information contained in the distributions remains unused for generating the generalized trajectory; and (3) the extraction of task constraints can only be performed at the level of the keypoints.

As for the *HMM continuous approach*, it presents two principal drawbacks: (1) the covariance information is not used to retrieve a generalized version of the trajectory, which is a severe loss of information since the continuous

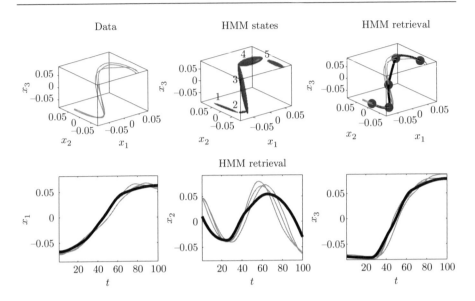

Figure 3.6 The encoding of continuous data (*thin grey line*) and the retrieval through interpolation (*thick black line*) when employing an HMM of 5 states (see also Table 3.2 and Fig. 3.5).

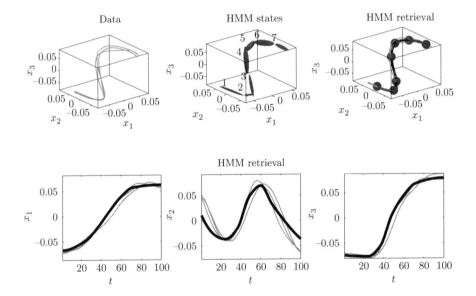

Figure 3.7 The encoding of the continuous data (*thin grey line*) and the retrieval through interpolation (*thick black line*) with an HMM of 7 states.

Table 3.2 The parameters of the HMM from the encoding
of raw data presented in Figure 3.6.

		Transition probability a_{ij}				
i	j	1	2	3	4	5
	1	**0.94**	**0.05**	0.00	0.00	0.00
	2	0.00	**0.93**	**0.07**	0.00	0.00
	3	0.00	0.00	**0.96**	**0.04**	0.00
	4	0.00	0.00	0.00	**0.95**	**0.05**
	5	0.00	0.00	0.00	0.00	**1.00**

Initial state probability Π_i				
1	2	3	4	5
1.00	0.00	0.00	0.00	0.00

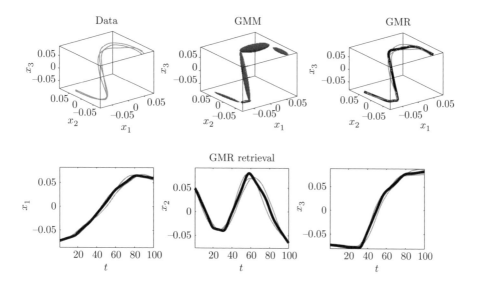

Figure 3.8 The encoding of the continuous data (*thin grey line*) and the retrieval
through GMR (*thick black line*) when using a GMM of 5 states.

trajectories considered encapsulate local correlation information directly in the
Gaussian distribution; and (2) the information contained in the two extremities
of the trajectories is lost during the reproduction process as only the center of
the Gaussian distribution is considered.

For this simple experiment, due to the analytic nature of the regression process, the combination of GMM and GMR to extract the constraints
of a task and to reproduce a generalized version of a set of trajectories is
clearly advantageous when compared to the HMM approaches considered here.

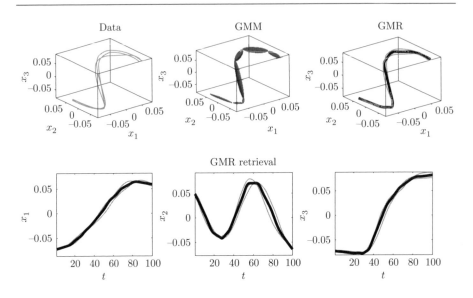

Figure 3.9 The encoding of the continuous data (thin grey line) and the retrieval through GMR (thick black line) when using a GMM of 7 states.

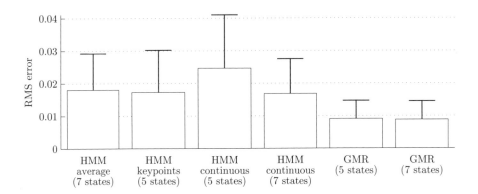

Figure 3.10 A comparison of reproduction efficiency for six reproduction processes based on HMM and GMM/GMR (see Figs. 3.3 and 3.9). A *root mean square* (RMS) error measure is used to compare the approaches when taking the original data as reference and by computing the RMS distance point by point with the generalized trajectory. The mean error and standard deviation for each method are depicted.

To control robots, the GMM/GMR approach allows the retrieval of the generalized signal at a desired rate and allows a control of the smoothness of the motion by appropriately selecting the number of Gaussians used in the model, e.g., to avoid vibrations that would cause important torques on the motors.

Table 3.3 A summary of the advantages/disadvantages of the reproduction methods (i.e., the *HMM keypoints* approach, the *HMM continuous* approaches and the *DTW/GMM/GMR* approach) from a theoretical and computational point of view.

	HMM keypoints	HMM continuous	DTW, GMM and GMR
Robustness to time distortions	★ (1)	★★★ (2)	★★ (3)
Robustness of the encoding process	★ (4)	★★	★★
Extraction of task constraints	★ (5)	★★ (6)	★★★ (7)
Reproduction capabilities	★★ (8)	★ (9)	★★★ (10)

(1)   The corresponding keypoints may be incorrectly grouped in various states of the HMM for strong temporal distortion across the demonstrations.

(2)   A double stochastic process to efficiently deal with temporal and spatial variations (but also more parameters, as compared to GMM, for a simultaneous learning).

(3)   The combination of a pattern-based approach, to deal with temporal variation (pre-processing by DTW), with a model-based approach, to deal with spatial variation (encoding through GMM). The main disadvantage over HMM is the tuning of several parameters for DTW.

(4)   The extraction of keypoints requires a pre-processing phase sensitive to the parameters of the segmentation process.

(5)   The constraints are extracted only for the keypoints previously extracted.

(6)   The variability and correlation information is represented only for each state of the HMM.

(7)   The variability and correlation information takes a continuous form and can be evaluated through GMR anywhere along the trajectory.

(8)   A generative model sensitive to the employed interpolation function.

(9)   A generative model sensitive to the employed interpolation function for which information at both extremities of the trajectory is lost.

(10)   A smooth reproduction using covariance information and simple handling of multiple constraints.

3.2   OPTIMAL LATENT SPACE OF MOTION

This experiment aims at comparing the use of *Principal Component Analysis* (PCA), *Canonical Correlation Analysis* (CCA) and *Independent Component Analysis* (ICA) (see Sect. 2.7) so as to find an optimal latent space of motion in order to linearly project motion data.

3.2.1   Experimental setup

For this experiment, the dataset consisted of multiple examples of 10 gestures inspired from the signals used by basketball officials to communicate various

Figure 3.11 A selection of 10 gestures used by basketball officials. Images reproduced from FIBA Central Board (2006) with permission.

non-verbal information such as scoring, clock-related events, administrative notifications, types of violations or types of fouls. These gestures are presented in Figure 3.11. Basketball officials' signals provide a rich gesture vocabulary characterized by non-linearities at the level of the joint angles, rendering them attractive as test data for researchers (Gross & Shi, 2001; Tian & Sclaroff, 2005).

The dataset consists of joint angles collected both by the user wearing motion sensors, as presented in Section 3.1 (3 demonstrations) and by a humanoid robot collecting kinesthetic data through manual movement of its arms (3 demonstrations). All the gestures start with the arms hanging along the body.

The experiment is conducted with an *HOAP-3* robot from the *Fujitsu* company, where *HOAP* stands for *Humanoid for Open Architecture Platform*.[7] This type of robot has 28 degrees of freedom (DOFs), of which only the 16 DOFs of the upper torso were used in this experiment (see Fig. 3.12). The remaining DOFs of the legs were set to a static posture.

The *kinesthetic teaching* process consisted in using motor encoders to record information while the teacher moved the robot's arms. The teacher selected the motors that he/she wanted to control manually by slightly moving them before the reproduction started. The selected motors were set to a passive mode, allowing the user to freely move the corresponding degrees of freedom while the robot executed the task. The interaction when embodying

---

[7] This robot is commercially available, see `http://www.automation.fujitsu.com/` for more information.

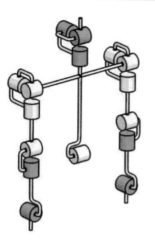

Figure 3.12 The degrees of freedom (DOFs) for the upper torso of the HOAP-3 robot. The backdrivable DOFs are represented in white and the DOFs that can only be moved by the robot are represented in grey.

the robot was more playful than the use of a graphical simulation would be and presented the advantage of the user implicitly feeling the robot's limitations in its real-world environment. This way, the teacher was able to provide partial demonstrations while the kinematics of each joint motion were recorded at a rate of 1000 Hz. The trajectories could then be resampled if required (each demonstration was here resampled to a fixed number of points $T = 100$). The robot was provided with motor encoders for every DOF, except for the hands and the head actuators (see Fig. 3.12).

With the motion sensors attached to the user's body, the robot could first *observe* the user performing one of the gestures. This motion was then improved by physically moving the robot's limbs while performing the gesture. The gesture was partially refined by grasping some of the robot's limbs and moving the desired DOFs while the robot controlled the remaining DOFs.[8] For each gesture, 3 demonstrations through motion sensors were first provided and this was followed by 3 demonstrations through kinesthetic teaching. Figure 3.13 illustrates the use of these two demonstration modalities.

3.2.2   Experimental results

Figure 3.14 and Table 3.4 present the results of an automatic estimation of the dimensionality, D, of the latent space of motion for the 10 gestures. Furthermore, Figure 3.15 gives an example for the projection in a latent space of motion when using either PCA, CCA and ICA. After using a GMM representation to

---

[8] Note that the number of variables that the user was able to control with his/her two arms was lower than when using the motion sensors. With kinesthetic teaching, the demonstrator could only control a subset of joint angles (the arms), while the robot controlled the remaining joints (the head and the hands).

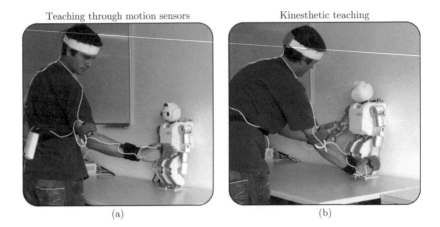

Figure 3.13 An illustration of the teaching modalities used in the experiment. (a) The user performs a demonstration of a gesture while wearing motion sensors recording his upper-body movements (arms and head). (b) The user helps the robot reproduce the gesture through kinesthetic teaching, i.e., he corrects the movement by physically positioning the robot limbs in their correct postures.

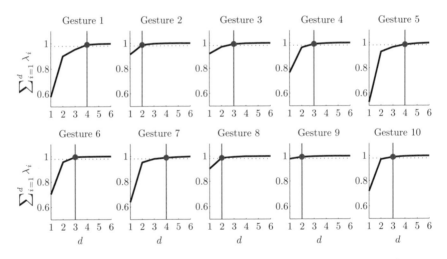

Figure 3.14 The automatic estimation of the dimensionality, D, of the latent space of motion for the 10 gestures used in the experiment. Here, d represents the candidate dimensionality and $\sum_{i=1}^{d} \lambda_i$ the cumulative sum of *eigenvalues*. For each graph, the selected dimensionality, D, was depicted by a point. The dashed line shows the threshold (at least 98% of the variance retained by the D first components).

model the data in this respective latent space, a generalized version of the data was computed through GMR (represented in solid, dotted, and dash-dotted lines). The generalized trajectories were then projected back in the original data space (see Fig. 3.16). The various criteria used to project the data gave rise to several projections. However, the criteria did not have a significant

Table 3.4 An automatic estimation of the dimensionality D of the latent space of motion for the 10 gestures used in the experiment.

Gesture ID	1	2	3	4	5	6	7	8	9	10
Dimensionality D	4	2	3	3	4	3	4	2	2	3

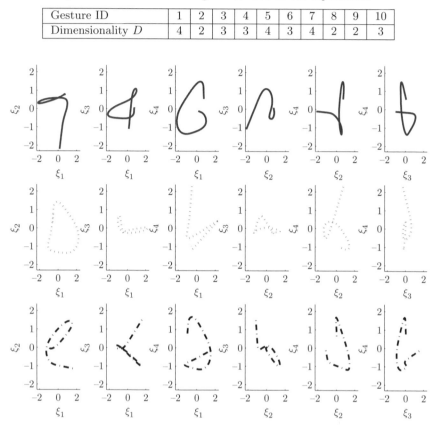

Figure 3.15 Generalized trajectories retrieved from the encoding of *Gesture 1* in a latent space of motion when using a linear transformation defined by PCA (*solid line*), CCA (*dotted line*) and ICA (*dash-dotted line*).

influence on the qualitativeness of the reconstructed data (the different re-projected data were superimposed). This could be quantitatively confirmed by the very small difference of $(1.2 \pm 0.5) \cdot 10^{-7}$ between the log-likelihood values of the GMMs computed in the original data space and those of the GMMs estimated either in the PCA, CCA and ICA latent space and projected back in the original data space. Based on this result, PCA was assumed to be a sufficient pre-processing step for further experiments.

The use of PCA and ICA to reduce the dimensionality of human motion data for a further encoding in HMM has also been investigated by Calinon and Billard (2005). In their experiments, recognition and reproduction performances were systematically evaluated for various ranges of noisy signals. The results demonstrated that the PCA-HMM and ICA-HMM combinations were both very successful at reducing the dimensionality of the dataset and

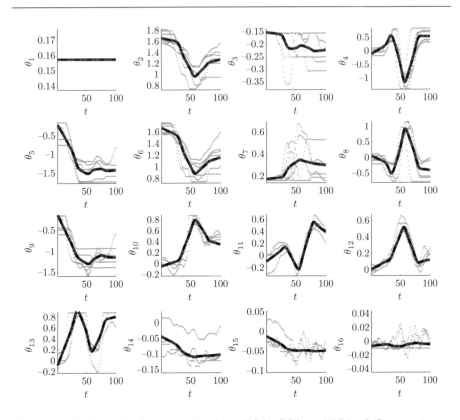

Figure 3.16 Generalized trajectories from PCA, CCA and ICA of *Gesture 1* projected back to the original data space (the reconstructed trajectories represented solid lines are superimposed).

at extracting the characteristics of each gesture. As expected, a preprocessing of the data using PCA and ICA removed the noise and increased the robustness of the HMM encoding. PCA/ICA also significantly reduced the amount of parameters required for encoding the gestures in the HMM. This experiment also led to the conclusion that ICA has very few advantages over PCA for the set of gestures considered. Even if the uncorrelatedness assumption of PCA was weaker than statistical independence, it constituted here a sufficient constraint to decompose simple human motion dataset. In other words, decorrelation appeared to be a sufficient pre-processing step when the data were further encoded in a GMM/HMM.

3.3 OPTIMAL SELECTION OF THE NUMBER OF GAUSSIANS

This experiment considers the automatic estimation of an optimal number of Gaussians, K, in the GMM.

3.3.1   Experimental setup

The experiment employed the same dataset of gestures inspired by basketball officials as the experiment presented in the previous section (see Fig. 3.11). The methods described in Sections 2.8.1 and 2.8.2 are considered to estimate the number of parameters required to encode a gesture in a GMM, based respectively on the *Bayesian Information Criterion* (BIC) and on the curvature analysis of the trajectory. The results were contrasted to a manual estimation of the number of Gaussians.

3.3.2   Experimental results

The estimation of the number of Gaussians required to model each of the 10 gestures is presented in Table 3.5. The BIC and curvature-based algorithms provided estimates that were close to those based on expert segmentation, with a difference of one component in most of the cases. BIC was found to be slightly better at estimating the number of parameters, but required the computation of multiple models. On the other hand, the curvature-based algorithm had the advantage of providing initial estimates for the GMM.

The automatic approximation of the number of Gaussians, K, by BIC is presented in Figure 3.17, and an example of the automatic estimation of K for *Gesture 1* by curvature-based analysis is provided in Figure 3.18. The figure also shows the initialization of the GMM parameters following this segmentation process, where the GMM parameters were updated by *Expectation-Maximization* (EM). For a majority of the gesturers, the curvature-based segmentation process gave rise to a qualitatively satisfying approximation for the initialization of the GMM parameters. Figure 3.19 displays a quantitative estimation based on log-likelihoods for measuring the quality of the initial estimates when using either the curvature-based segmentation method or the *k-means* initialization process (with the same number of Gaussians extracted by curvature-based segmentation). In most of the cases, *k-means* were able to give the best initial estimate (in terms of log-likelihood) of the parameters, which was often very close to the local optimum (EM only slightly increased the log-likelihood).

The initialization obtained from curvature-based segmentation still provided decent estimates that could then be refined by EM. It thus remained an interesting alternative to the *k-means* initialization scheme as it permitted

Table 3.5 An estimation of the number of Gaussians, K, in the GMM for the 10 gestures used in the experiment when employing either *Bayesian Information Criterion* (BIC), a curvature-based method or a manual segmentation from an expert. Only the first demonstration was used to select the optimal number of components.

Gesture ID	1	2	3	4	5	6	7	8	9	10
BIC	4	4	5	4	3	4	4	4	4	5
Curvature-based	5	3	3	6	6	6	3	3	3	3
Manual segmentation	5	4	4	5	6	6	4	4	4	4

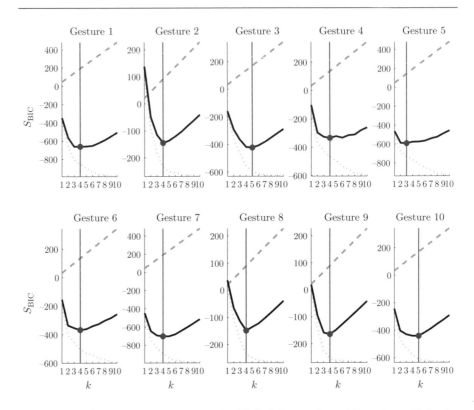

Figure 3.17 An automatic estimation by BIC of the number of Gaussians, K, in the GMM for the 10 gestures used in the experiment. For each graph, the dotted line represents the inverse log-likelihood, $-\mathcal{L}$, the dashed line corresponds to $\frac{n_p}{2}\log(N)$, and the solid line gives the final score S_{BIC}. The point represents the selected number of components (minimum of S_{BIC}).

a simultaneous initialization of the GMM and estimation of the number of parameters required to model the gesture (there was no need to evaluate multiple models through BIC). Despite these promising results, BIC will be used in further experiments to encode gestures in GMMs as a result of it producing slightly better approximations of the number of Gaussians.

3.4 ROBUSTNESS EVALUATION OF THE INCREMENTAL LEARNING PROCESS

This experiment evaluated the robustness of the two incremental teaching procedures presented in Section 2.6.3. Their performances were contrasted with a batch learning process based on *Expectation-Maximization* (EM).

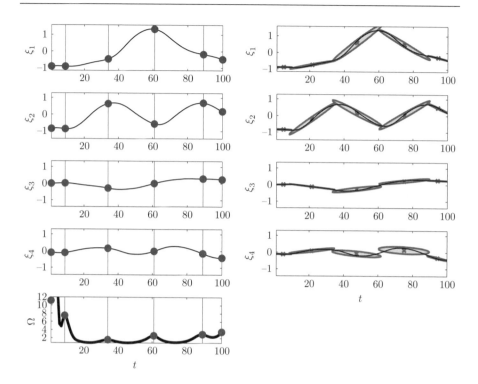

Figure 3.18 The segmentation of *Gesture 1* by the curvature-based method (*left column*), where the segmentation points are used to initialize the GMM parameters (*right column*). Local maxima of the trajectory Ω representing the curvature (*last row*) were used to segment the trajectory ξ.

3.4.1   Experimental setup

The basketball officials' dataset presented in Section 3.1 (see Fig. 3.11) was used also in this experiment, and two modalities were selected to convey the demonstrations (see Fig. 3.13).

PCA was employed as a pre-processing step to project the original joint angle trajectories into a subspace of reduced dimensionality (see Sect. 3.2) and the *Bayesian Information Criterion* (BIC) was used to select an optimal number of Gaussians in the GMM (see Sect. 3.3).[9] The original data space of 16 DOFs was then reduced to a latent space of 2-4 dimensions, where 3-5 Gaussians per GMM were required to encode the gestures.

The efficiency of the *Batch* (B), *Incremental – direct update* (IA) and *Incremental – generative* (IB) learning processes, were estimated by computing the log-likelihood of the resulting model when tested with the training set.

---

[9] Note that only the first demonstration was used to determine the optimal number of components as the gesture is incrementally learned.

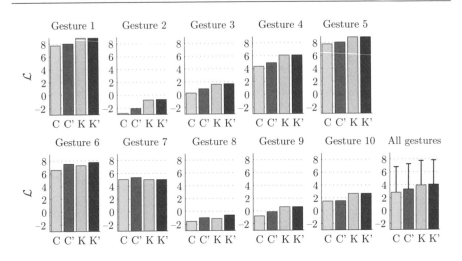

Figure 3.19 A comparison of the log-likelihood, $\mathcal{L}$, when initialized with a curvature-based segmentation process as opposed to with *k-means* (using the same number of components). C and C' define respectively the log-likelihood, $\mathcal{L}$, at initialization (using a curvature-based process) and after EM convergence. Correspondingly, K and K' define respectively the log-likelihood, $\mathcal{L}$, at initialization (using *k-means*) and after EM convergence. The results are presented separately for each gesture and the last graph gives the average results with standard deviations.

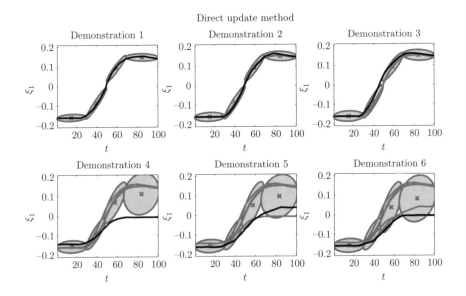

Figure 3.20 An illustration of the incremental learning process when using the direct update method (see Sect. 2.6.3), which is a modified version of the EM algorithm for incremental update of the GMM parameters. The algorithm takes into account only the novel observation to update the model (*represented with a black line*). Only the first variable in latent space of *Gesture 2* is represented here.

3.4.2   Experimental results

Figure 3.20 shows an example of the encoding results of GMM when incrementally updating the parameters with the *direct update method* presented in Section 2.6.3. The first row shows the three demonstrations provided when employing motion sensors and the second row portrays the three demonstrations obtained through kinesthetic teaching. The incremental training process efficiently refines the model after each demonstration without using historical data to update the model.

Computations of the log-likelihoods of the model when faced with the various demonstrations were carried out for each of the 10 gestures and at each of the 6 iterations (gestures already observed and further gestures to observe). The results averaged over the 10 gestures are reported in Figure 3.21, where *Model i* refers to the model after the *i*-th iteration and *Data i* refers to

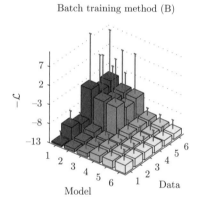

Figure 3.21 The evolution of the inverse log-likelihood, $-\mathcal{L}$, when trained with the three methods. After each demonstration, the current model was compared to the whole dataset (i.e., data already observed but also further data). The results were averaged over the 10 gestures. The 3D vertical bars and vertical segments represent means and standard deviations.

the i-th demonstration. Following the first demonstration, *Model 1* gives a very good description of *Data 1* (the log-likelihood was high, and $-\mathcal{L}$ was thus low). This first model also partially described *Data 2-3*, but gave a poor representation of *Data 4-6*. From the fourth demonstration, *Model 4* started to properly depict the whole training set, which was finally refined in *Model 6*. The various training methods showed no significant differences. Particularly, after the sixth demonstration, the log-likelihoods of *Model 6* trained with one of the incremental methods were almost constant for *Data 1-6*. Thus, the model was able to encode the previously observed data as well as those newly observed.

For each gesture, the log-likelihoods of the final model (*Model 6*) is reported in Figure 3.22, in which it can be seen that the resulting GMM representations for the incremental training methods IA and IB provided very similar likelihoods to the batch learning method B. Both direct update and generative methods were thus efficient for refining the GMM representation of the data. The loss of performance induced by the incremental training procedures was negligible as compared to the benefit of the methodology. Indeed, a learning system capable of forgetting historical data is advantageous in terms of memory capacity and scaling-up issues. Figure 3.23 presents reproduction results where the essential characteristics of the motion were correctly retrieved by the various incrementally trained models.

For the dataset considered in this experiment, the two incremental approaches demonstrated both advantages and drawbacks. The direct update approach did not rely on a stochastic process for updating the model. For

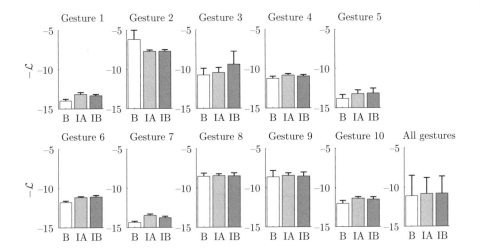

Figure 3.22 The inverse log-likelihood, $-\mathcal{L}$, of the models after observation of the sixth demonstration and when faced with the six demonstrations used to train the corresponding models. The results correspond to the ten gestures and the three teaching approaches, where B stands for the batch training procedure, and IA and IB relate respectively to the *direct update method* and the *generative method*. The last graph presents the results averaged over the entire set of gestures.

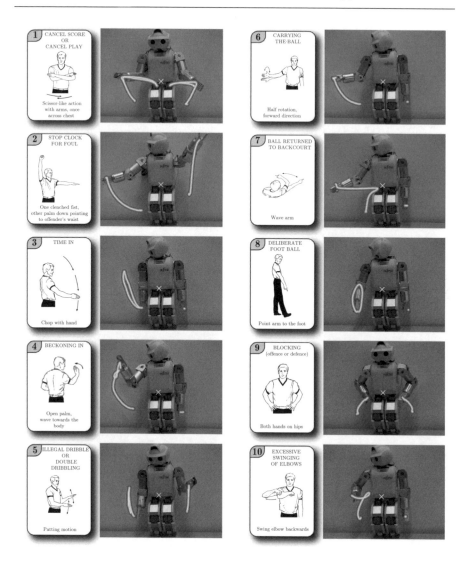

Figure 3.23 A reproduction of the 10 gestures by using an incremental training process (generative method) after having observed 6 demonstrations for each gesture. The trajectories of the hands were retrieved by projecting back in joint space the generalized trajectories encoded in latent space and by executing the resulting joint angle trajectories on the robot.

large datasets, this approach could thus represent advantages in terms of computation time.[10] However, the dataset is required to exhibit slow modification properties between two consecutive examples, due to the direct update method

---

[10] For the proposed experiment, the differences in computation time between the procedures are not very relevant as all the processes run in less than 0.1 sec. using a Matlab implementation on a standard PC.

assuming that the cumulated posterior probability does not change much with the inclusion of the novel data in the model. This constraint is partially fulfilled when considering data that are first recognized by the model (i.e., sharing enough similarities with the previous data) before updating the model parameters. The generative update method may be more robust when considering datasets presenting a high variability between two consecutive demonstrations. This issue is discussed further in Section 8.2.2.

CHAPTER 4

HANDLING OF CONSTRAINTS IN JOINT SPACE AND TASK SPACE

This chapter presents experiments to automatically extract the important characteristic of a skill and determine an appropriate controller for reproduction considering constraints in task space and joint space. It illustrates the use of the proposed probabilistic framework in HRI experiments by gradually increasing the complexity of the system in a didactic manner. The inverse kinematics issue is first introduced, after which its integration into a probabilistic framework is discussed.

4.1 INVERSE KINEMATICS

A robotic arm with n degrees of freedom (DOFs) manipulating objects in a m-dimensional space is considered. The arm posture and the end-effector location are given by the vectors $\theta \in \mathbb{R}^n$ and $x \in \mathbb{R}^m$, respectively.[1] These vectors are indexed by time t, indicating that the arm is in joint configuration θ_t with the end-effector at location x_t at time t. We define $x = K(\theta)$ as the kinematic function determined by the geometry of the robotic arm and is assumed to be known. With redundant robotic arms, several values of θ may yield the same value $K(\theta)$.

Inverse kinematics concerns the problem of finding a set of joint angles θ to follow a desired trajectory x in Cartesian space. Local inverse kinematics solutions are here considered. The problem can be stated as follows. The manipulator is in joint configuration θ, and one would like to modify its joint angles in order for its end-effector to move short distances in a desired direction in Cartesian space. The aim is then to find the joint angle velocities $\dot{\theta}$ corresponding to a desired end-effector velocity $\dot{x}$. One seeks a $\dot{\theta}$ such that

$$K(\theta + \tau\dot{\theta}) = K(\theta) + \tau\dot{x},$$

---

[1] More generally, x describes the pose of the end-effector, for instance represented by position, orientation or both.

where τ is a time constant that tends to zero. Deriving with respect to the time τ yields

$$J(\theta)\dot{\theta} = \dot{x} \quad \text{with} \quad J(\theta) = \frac{\partial}{\partial\theta}K\mid_\theta, \tag{4.1}$$

where the matrix $J(\theta)$ is called the Jacobian of the kinematic function K. The local inverse kinematics problem consists in solving (4.1) with respect to $\dot{\theta}$. If the matrix $J(\theta)$ is square and invertible (i.e. $n = m$), the solution is simply $\dot{\theta} = J(\theta)^{-1}\dot{x}$. In this case, (4.1) has a unique solution. However, if $\operatorname{rank}(J(\theta)) < n$, there is no solution to (4.1) and if $\operatorname{rank}(J(\theta)) > n$, there is an infinite number of solutions. In the latter case, all solutions belong to an affine space known as the *solution space* and the equation $J(\theta)\dot{\theta} = 0$ defines a vector space denoted the *null space*. Adding any vector $\dot{\theta}^n$ of the null space to a vector $\dot{\theta}^s$ of the solution space yields a vector belonging to the solution space, i.e.,

$$J \cdot (\dot{\theta}^n + \dot{\theta}^s) = J\dot{\theta}^n + J\dot{\theta}^s = 0 + \dot{x} = \dot{x}.$$

In the above equation as well as in those forthcoming, the dependency of J on θ will sometimes be dropped in order to lighten the notation. The problem is now to find an optimal controller among all $\dot{\theta}$ belonging to the solution space. In the following sections, the most popular solutions to this problem are briefly described (see Hersch (2008) and Sciavicco, Siciliano, and Sciavicco (2000) for more exhaustive reviews of the inverse kinematics solutions).

4.1.1   Local solutions

Least Norm (pseudoinverse) method
The first solution to this problem was suggested in the late sixties by Whitney (1969) and involve taking the smallest $\dot{\theta}$ (least-norm solution). This appears to be an energy-efficient solution as one moves as little as possible to attain the desired end-effector velocity. In other words, the resulting posture at each iteration constitutes the one closest to the actual posture θ while lying on the solution space. This solution is obtained by solving the constrained problem[2]

$$\min_{\dot{\theta}} \|\dot{\theta}\|^2 \quad \text{u.c.} \quad J\dot{\theta} = \dot{x}. \tag{4.2}$$

This is easily done with the Lagrange multipliers technique, yielding

$$\dot{\theta} = J^\top (JJ^\top)^{-1} \dot{x} = J^\dagger \dot{x}.$$

Here, $J^\dagger$ is called the *Moore-Penrose pseudoinverse* (valid for redundant robot kinematics) and satisfies $JJ^\dagger = I$, where I is the identity matrix. The resulting position is simply the orthogonal projection of θ on the solution space.

This solution is however not always applicable as it is sometimes the case that the matrix $JJ^\top$ is not invertible, i.e., singular. Such circumstances correspond to particular values of θ, called *singular configurations*, that occur

---

[2] Here, u.c. refers to *under constraint*.

when the tangent plane is orthogonal to a subspace of x, which means that there are directions in which the robot cannot move. Moreover, this method does not allow joint limits to be avoided and is therefore likely to bring the robot out of its motion ranges.

Weighted Least Norm method

The Least Norm method has been extended to the Weighted Least Norm (WLN) method in order to provide the possibility of weighing the importance of the different joints for the motion (Whitney, 1972). This leads to particular joints being moved as little as possible while giving more freedom to other joints, which can be expressed by the following constrained minimization problem

$$\min_{\dot\theta} \quad \dot\theta^{\mathsf T} W^\theta \dot\theta \quad \text{u.c.} \quad J\dot\theta = \dot x.$$

Here, W^θ is a square diagonal matrix of size n. Numbers on the diagonal indicate the "costs" of moving the corresponding joint. This problem is also solved with the Lagrange multipliers technique, yielding

$$\dot\theta = W^\theta J^{\mathsf T} (J W^\theta J^{\mathsf T})^{-1} \, \dot x. \tag{4.3}$$

which can be seen as a Least Norm solution in a "stretched" joint angle space, where the stretching factors along the joint angle dimensions are given by the square root of the corresponding elements of W^θ. As such, it is also vulnerable to singularities, but joint limits may still be avoided by providing weights according to the proximity of each joint angle to their corresponding limits. The reader is referred to Chan and Dubey (1995) for details.

Gradient Projection method

This method has been suggested by Liegeois (1977) for addressing the joint limit avoidance problem. The idea is to take advantage of the null space of the Jacobian to optimize a cost function bringing the joint angles to the center of their range, which can be expressed by the optimization problem

$$\min_{\dot\theta} \quad \dot\theta^{\mathsf T} \dot\theta + \lambda \, \mathcal{H}(\theta + \tau\dot\theta) \quad \text{u.c.} \quad J\dot\theta = \dot x.$$

Here, $\mathcal{H}$ is the function to optimize, and $\lambda \in \mathbb{R}$ weighs the influence of the optimization function with respect to the joint velocity. The solution to this problem is given by

$$\dot\theta = J^\dagger \dot x + \lambda (I - J^\dagger J)\nabla \mathcal{H}, \tag{4.4}$$

where ∇ is the gradient operator. The first term corresponds to the Least Norm solution and the second term is the projection of the gradient of $\mathcal{H}$ in the null space of J so that it does not influence $J\dot\theta$. If $\mathcal{H}$ is used for joint limit avoidance, a typical expression for it is

$$\mathcal{H}(\theta) = \frac{1}{n}\sum \left(\frac{\theta_i - \bar\theta_i}{\theta_i^M - \bar\theta_i} \right)^2 \quad \text{with} \quad \bar\theta_i = \frac{1}{2}(\theta_i^m + \theta_i^M),$$

where θ_i^m, $\bar\theta_i$ and θ_i^M are respectively the lower boundary, mid-range and upper boundary of joint angle i. More generically, if $g(\theta)$ is the gradient of the cost function to optimize, we have

$$\dot{\theta} = \overbrace{J^{\dagger}\dot{x}}^{\dot{\theta}^{s}} + \overbrace{\lambda(I - J^{\dagger}J)g(\theta)}^{\dot{\theta}^{n}},$$

where $\dot{\theta}^{s}$ is the velocity component in solution space and $\dot{\theta}^{n}$ is the velocity component in null space (also called homogeneous solution). We see that, regardless of the function $g(\theta)$, the velocity component $\dot{\theta}^{n}$ of the controller does not modify the velocity in Cartesian space

$$\dot{x}^{n} = J\dot{\theta}^{n} = \lambda(J - JJ^{\dagger}J)g(\theta) = \lambda(J - \overbrace{JJ^{\top}(JJ^{\top})^{-1}}^{I}J)g(\theta) = 0.$$

Damped Least Squares method

The Damped Least Squares method was proposed simultaneously by Wampler (1986) and Nakamura and Hanafusa (1986) to deal with singularities. As mentioned above, singularities occur when the robot is incapable of moving in a certain direction due to the configuration of its joints. In this case, the constraint given in (4.2) cannot be met and the problem is without solution. To solve this issue, an approximate solution to the optimization problem can be defined as

$$\min_{\dot{\theta}} \quad \|J\dot{\theta} - \dot{x}\|^{2} + \lambda\|\dot{\theta}^{2}\| \quad \text{with} \quad \lambda > 0,$$

which leads to

$$\dot{\theta} = (J^{\top}J + \lambda I)^{-1}J^{\top}\dot{x}.$$

A weighted expression of the Damped Least Squares (DLS) can also be defined as

$$\min_{\dot{\theta}} \quad (J\dot{\theta} - \dot{x})^{T}W^{x}(J\dot{\theta} - \dot{x}) + \dot{\theta}^{\top}W^{\theta}\dot{\theta},$$

giving rise to the solution

$$\dot{\theta} = (J^{\top}W^{x}J + W^{\theta})^{-1}J^{\top}W^{x}\dot{x}, \qquad (4.5)$$

where W^{x} and W^{θ} are diagonal positive definite matrices of size m and n. Several variations of this method are detailed in Chiaverini, Oriolo, and Walker (2008).

4.1.2　Extending inverse kinematics solutions to a statistical framework

We have seen in Section 2.5 how to combine different constraints in a statistical framework by considering that the various models were in the same representation space (e.g., by combining constraints with respect to a variety of objects in Cartesian space). We now consider the more generic case where we seek an optimal controller that combines constraints in joint space and task space.

To illustrate this issue, let us first consider the toy example of a planar manipulator of two DOFs performing two demonstrations of a task by starting from varying joint angle configurations (Fig. 4.1a). By observing the joint angle trajectories $\{\theta_1, \theta_2\}$ and the paths followed by the end-effector $\{x_1, x_2\}$, we see that the first set of variables changed significantly while the second set of variables remained fairly similar across the demonstrations. Thus, the information conveyed by the end-effector path appears to be more reliable than that conveyed by the joint angles. In other words, the task seems to put stronger constraints on the end-effector path than on the joint angle trajectories. As a result, in order to reproduce the task, one would give more weight to reproducing the end-effector path than the joint angle trajectories. Once measuring the relative importance of each set of variables and incorporating this measure into a cost function $\mathcal{H}$, an optimal controller with respect to $\mathcal{H}$ can be computed. The problem may not be as straightforward as it seems since the initial situation may vary, and/or the demonstrator and the imitator may deviate significantly in their embodiment (arm lengths, number of degrees of freedom for each arm). This issue is illustrated in Figure 4.1, where the two segments of the imitator's manipulator differ in length from those of the demonstrator. If the imitator's arm replays directly the joint angle trajectories of the demonstrator's arm, we would end up with a very different path for the end-effector than the one demonstrated. Conversely, replaying the end-effector path (using for example a least-norm inverse kinematics controller) would result in very different joint angle trajectories. This figure illustrates the effect of generating several alternatives, from purely satisfying the joint trajectories to satisfying only the end-effector path.

A similar issue is considered when transferring a skill from a human to a robot, i.e., when considering either another embodiment or a variety of objects to manipulate (see Fig. 1.2). Figure 4.2 provides an example for a *Chess Task*

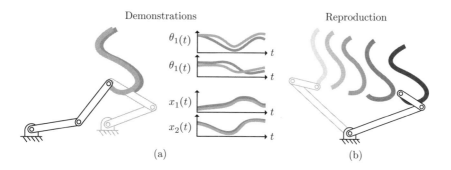

Figure 4.1 (a) An illustrative scheme of the *what-to-imitate* issue. A 2-DOF manipulator arm produces two demonstrations of a given task, namely drawing an S figure, starting from various joint angle configurations. The path of the end-effector (given by $\{x_1, x_2\}$) is invariant, while the joint trajectories $\{\theta_1, \theta_2\}$ vary significantly. (b) An illustrative scheme of the *how-to-imitate* issue. As the manipulator considered for reproduction has a different embodiment, it can generate several alternatives to reproduce the demonstrated task, from purely satisfying the joint angle trajectories (a) to satisfying only the hand path (b). The issue is then to find an appropriate trade-off between satisfying the constraints related respectively to the constraints in joint space and task space.

Demonstrations   Reproduction

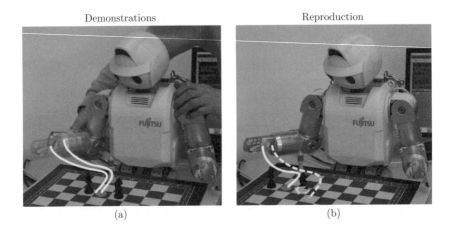

(a)   (b)

Figure 4.2 (a) An illustration of the *what-to-imitate* issue. (b) An illustration of the *how-to-imitate* issue.

consisting of grabbing a chess piece and moving it two squares forward. The left snapshot shows the path taken by the robot's hand during kinesthetic teaching when starting from two initial locations. To extract the constraints of the task (i.e., to determine *what-to-imitate*), the robot computes the variations and correlations across the variables. In this task, such an analysis reveals weak correlations at the beginning of the motion, as a large set of paths can be used to reach for the chess piece depending on the initial position of the hand. However, the analysis reveals a strong correlation for grabbing the piece and pushing it towards the desired location without hitting the other chess pieces on the chessboard. The right snapshot of Figure 4.2 illustrates the *how-to-imitate* issue. Once trained to perform a task in a particular context, the robot should be able to generalize and reproduce the skill in a different situation. In this example, the robot should manage to grab and move the chess piece two squares forward wherever it may be on the chess board. Since the demonstrated joint angles and hand path are mutually exclusive in the imitator space, it is not possible to fulfill both constraints simultaneously. Depending on the situation, the robot may have to find very different joint angle configurations as compared to the one demonstrated. To this end, the robot computes the controller that gives the optimal trade-off between satisfying the constraints of the task and its own body constraints. The following sections present a set of experiments to demonstrate how this trade-off can be computed.

4.2 HANDLING OF TASK CONSTRAINTS IN JOINT SPACE – *EXPERIMENT WITH INDUSTRIAL ROBOT*

To treat both end-effector paths x and joint angles θ in the same representational space, velocities are first estimated from the set of trajectories collected by the robot as

$$\dot{x}_t = \frac{1}{\Delta t}(x_t - x_{t-\Delta t}),$$

$$\dot{\theta}_t = \frac{1}{\Delta t}(\theta_t - \theta_{t-\Delta t}),$$

where t refers to the t-th time step elapsed from the beginning of the demonstration.

$\dot{\theta}$ and $\dot{x}$ are kinematically constrained as discussed in Section 4.1. The generalized joint angle velocities $\hat{\dot{\theta}}$ and generalized end-effector velocities $\hat{\dot{x}}$ (both retrieved through the GMM/GMR approach presented in Sect. 2.4) can still be mutually exclusive in the imitator workspace, due to the different embodiment between the demonstrator and the imitator or to the varying situations between demonstrations and reproduction attempt, or simply because the generalization processes are performed separately. To simultaneously fulfill constraints in joint space and task space, the product properties of normal distributions (see Eq. (2.6) in Sect. 2.5.2) can be used together with local inverse kinematics properties to find a controller satisfying the two constraints.

Let $\hat{\dot{\theta}}$ and $\hat{\dot{x}}$ be respectively the generalized joint angle velocities and the generalized end-effector velocities in Cartesian space. Let $\dot{\theta}$ and $\dot{x}$ be the candidate velocities for reproducing the motion. An optimal controller can be determined by solving the constrained optimization problem

$$\min_{\dot{\theta},\dot{x}} \quad (\dot{\theta} - \hat{\dot{\theta}})^\top W^\theta (\dot{\theta} - \hat{\dot{\theta}}) + (\dot{x} - \hat{\dot{x}})^\top W^x (\dot{x} - \hat{\dot{x}})$$

$$\text{u.c.} \quad \dot{x} = J\dot{\theta}, \tag{4.6}$$

where J is the Jacobian matrix at posture θ_t. $W^\theta \in \mathbb{R}^{n \times n}$ and $W^x \in \mathbb{R}^{m \times m}$ are semi-definite positive diagonal matrices serving as coefficients indicating the respective influence that one should give to the desired joint angles and end-effector location. Coherence is then enforced by finding the coherent velocities $(\dot{x}, \dot{\theta})$ that will bring the system closest to $(\hat{\dot{x}}, \hat{\dot{\theta}})$.

A solution can be obtained using Lagrange multipliers. The quantity to minimize can be expressed as

$$L(\dot{\theta}, \dot{x}) = (\dot{\theta} - \hat{\dot{\theta}})^\top W^\theta (\dot{\theta} - \hat{\dot{\theta}}) + (\dot{x} - \hat{\dot{x}})^\top W^x (\dot{x} - \hat{\dot{x}}) - \lambda^\top (\dot{x} - J\dot{\theta}),$$

where λ is the vector of Lagrange multipliers. After differentiating with respect to $\dot{\theta}$ and $\dot{x}$, and setting the equation to zero, one gets

$$\frac{\partial L}{\partial \dot{\theta}} = 2W^\theta(\dot{\theta} - \hat{\dot{\theta}}) + J^\top \lambda = 0, \tag{4.7}$$

$$\frac{\partial L}{\partial \dot{x}} = 2W^x(\dot{x} - \hat{\dot{x}}) - \lambda = 0. \tag{4.8}$$

Combining (4.7) and (4.8) yields

$$W^\theta(\dot{\theta} - \hat{\dot{\theta}}) + J^\top W^x(\dot{x} - \hat{\dot{x}}) = 0,$$

and solving for $\dot{\theta}$ gives

$$\dot{\theta} = (W^{\theta} + J^{\top}W^{x}J)^{-1}(J^{\top}W^{x}\hat{\dot{x}} + W^{\theta}\hat{\dot{\theta}}). \tag{4.9}$$

An alternate representation of (4.9) can be formulated by first rewriting (4.7) and (4.8) as

$$\dot{\theta} = \hat{\dot{\theta}} + \frac{1}{2}(W^{\theta})^{-1}J^{\top}\lambda, \tag{4.10}$$

$$\dot{x} = \hat{\dot{x}} - \frac{1}{2}(W^{x})^{-1}\lambda. \tag{4.11}$$

Inserting these equations into (4.6) then yields

$$J\left(\hat{\dot{\theta}} + \frac{1}{2}(W^{\theta})^{-1}J^{\top}\lambda\right) = \hat{\dot{x}} - \frac{1}{2}(W^{x})^{-1}\lambda,$$

$$\lambda = 2\left((W^{x})^{-1} + J(W^{\theta})^{-1}J^{\top}\right)^{-1}(\hat{\dot{x}} - J\hat{\dot{\theta}}),$$

which is subsequently used to replace λ in (4.10)

$$\dot{\theta} = \hat{\dot{\theta}} + (W^{\theta})^{-1}J^{\top}\left((W^{x})^{-1} + J(W^{\theta})^{-1}J^{\top}\right)^{-1}(\hat{\dot{x}} - J\hat{\dot{\theta}}). \tag{4.12}$$

Although simpler than (4.12), (4.9) may be disadvantageous from an implementation perspective. Indeed, using $(W^{\theta})^{-1}$ and $(W^{x})^{-1}$ instead of W^{θ} and W^{x} renders it possible to avoid infinity measures when dealing with purely angular controllers. It is indeed easier to handle $(W^{\theta})^{-1}$ (or $(W^{x})^{-1}$) as equal to zero during computation. Moreover, this formulation is faster as it requires a matrix inversion of degree m, whereas (4.9) requires the inversion of a matrix of degree $n > m$ (for redundant manipulators). The parameters W^{θ} and W^{x} control the influence of each of the sub-controllers. By setting W^{x} to zero, one obtains a pure joint angle controller and by setting W^{θ} to zero, the result becomes a pure end-effector location controller.

The results given by (4.12) or (4.9) are in fact generalizations of the *Weighted Least Norm* solution (4.3) and of the *Weighted Damped Least Square* solution (4.5) described in Section 4.1. Indeed, by setting $\hat{\dot{\theta}} = 0$ (least-norm controller) and $(W^{x})^{-1} = 0$ in (4.12), one obtains the Weighted Least Norm method described in (4.3), which makes use of the Moore-Penrose pseudoinverse of the Jacobian to compute the adequate joint angle velocities. Furthermore, by setting $\hat{\dot{\theta}} = 0$ in (4.9), one obtains the Weighted Damped Least Square solution described in (4.5) to avoid singularities (Hersch, 2008).

Instead of analytically deriving a metric of imitation, the same result can be retrieved through the product property of normal distributions. Let us assume that the constraints in task space are defined by $x \sim \mathcal{N}(\hat{\dot{x}}, \hat{\Sigma}^{\dot{x}})$, and that the constraints in joint space are defined by $\dot{\theta} \sim \mathcal{N}(\hat{\dot{\theta}}, \hat{\Sigma}^{\dot{\theta}})$. When using the linear transformation property of normal distributions (2.5) and the inverse kinematics relation $\dot{\theta} = J^{\dagger}\dot{x}$, the constraints in task space can be represented in joint space as

$$\dot{\theta}' \sim \mathcal{N}(J^{\dagger}\hat{\dot{x}}, J^{\dagger}\hat{\Sigma}^{\dot{x}}(J^{\dagger})^{\top}). \qquad (4.13)$$

Based on the product property of normal distributions (2.6), the product of the two Gaussian distributions describing $\dot{\theta}$ and $\dot{\theta}'$ is equivalent (up to a scaling factor) to a Gaussian distribution $\mathcal{N}(\hat{\dot{\theta}}, \hat{\Sigma}^{\dot{\theta}})$ with parameters[3]

$$\dot{\theta} = \left((\hat{\Sigma}^{\dot{\theta}})^{-1} + J^{\top}(\hat{\Sigma}^{\dot{x}})^{-1}J\right)^{-1} \left((\hat{\Sigma}^{\dot{\theta}})^{-1}\hat{\dot{\theta}} + J^{\top}(\hat{\Sigma}^{\dot{x}})^{-1}\hat{\dot{x}}\right),$$

$$\Sigma^{\dot{\theta}} = \left((\hat{\Sigma}^{\dot{\theta}})^{-1} + J^{\top}(\hat{\Sigma}^{\dot{x}})^{-1}J\right)^{-1}. \qquad (4.14)$$

By considering $W^{\theta} = (\hat{\Sigma}^{\dot{\theta}})^{-1}$ and $W^{x} = (\hat{\Sigma}^{\dot{x}})^{-1}$, we then see that (4.9) and (4.14) provide the same result. Compared to the derivation of a cost function through Lagrange multipliers, the direct computation method has the advantage of not only determining an optimal controller satisfying several constraints but also to compute the resulting constraints for this controller, represented as $\hat{\Sigma}$ in (4.14). It should be noted that, instead of considering $\dot{x} = J\dot{\theta}$ as the coherence constraint, it is possible to define a coherence constraint that takes advantages of the robot kinematic redundancy. One can, for instance, consider an optimization term that does not modify the coherence constraint through projection in the null space of the Jacobian.

4.2.1   Experimental setup

The setup of the experiment is presented in Figure 4.3. Two 5-DOF *Katana* robots from *Neuronics*, characterized by a repeatability of ±0.1 mm and a maximum speed of $68°\,\text{sec}^{-1}$, were used for the experiment. A sixth motor controls the opening and closing status of the gripper, which is generated through a binary signal generalized over multiple demonstrations as described in Section 2.11.1. Each motor is equipped with encoders permitting the user to move the robot manually while registering joint angle information (see Fig. 4.3). During this process, the position of the end-effector is also computed through direct kinematics.

A stereoscopic vision system based on two webcams of 320×240 pixels was used to track a set of objects in 3D Cartesian space based on tracking in $YCbCr$ color space of colored patches attached to the objects (only Cb and Cr were robust to changes in luminosity), where each object to track was predefined in a calibration phase. The images were processed at a frame rate of 15 Hz by the *OpenCV* vision processing software, where each object was pre-defined in a calibration phase by fitting a Gaussian distribution on the $CbCr$ subspace

---

[3] Note that, for redundant robot kinematics, $JJ^{\dagger} = I$ and $J^{\dagger}\hat{\Sigma}^{\dot{x}}(J^{\dagger})^{\top}$ is not full rank. In this situation, a pseudoinverse needs to be used instead of an inverse, yielding $\left(J^{\dagger}\hat{\Sigma}^{\dot{x}}(J^{\dagger})^{\top}\right)^{\dagger} = ((J^{\dagger})^{\top})^{\dagger}(\hat{\Sigma}^{\dot{x}})^{\dagger}(J^{\dagger})^{\dagger} = J^{\top}(\hat{\Sigma}^{\dot{x}})^{-1}J$.

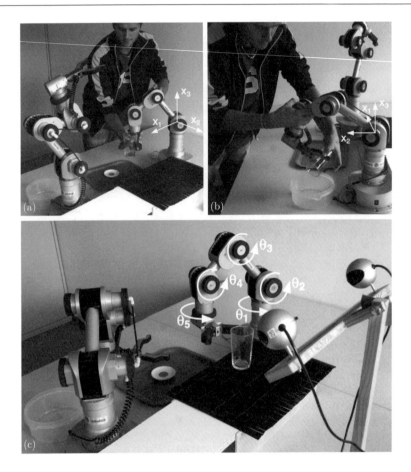

Figure 4.3 (a) Kinesthetic demonstrations of the two tasks considered in the experiment, namely grasping and placing a glass on a coaster, and grasping and emptying a glass (b). (c) A reproduction of the skill by the two robots, during which the new positions of the objects were tracked by a stereoscopic vision system.

characterizing the color of the object. Then, tracking was performed by using an adaptive search window similar to the *Continuously Adaptive Mean Shift* (CamShift) algorithm (Bradski, 1998). To calibrate the webcams, the intrinsic and extrinsic parameters were estimated according to the method proposed by Z. Zhang (2000).

Two skills were considered in the experiment, namely setting the table by grasping a glass on a shelf and placing it on a coaster, and clearing the table by grasping the glass from the table and emptying the glass in a basin (see Fig. 4.3). For the first task, two objects were tracked by the robot (the glass and the coaster), where the positions of the two objects could vary. For the second task, only one object was tracked by the robot, i.e., the glass was assumed to cover the coaster and the basin was assumed to have a fixed position in the robot's workspace.

The collected dataset was composed of the joint angle trajectories of the robot θ and the position of the end-effector x in the Cartesian space with respect to the objects detected in the scene. For the first task, five demonstrations of 11-dimensional trajectories were collected (5 variables describing the joint angles and 2×3 variables portraying the relative position of the end-effector with respect to the two objects), where each trajectory consisted of 1000 datapoints. For the second task, only 8-dimensional trajectories were considered due to only a single object being used.

By employing the GMM/GMR method presented in Section 2.4, the constraints in task space were computed with respect to the objects detected by the robot in its environment. The constraints associated with the position of the end-effector in the frame of reference of an object n were thus represented by the trajectories $\hat{x}^{(n)}$ and associated covariance matrices $\hat{\Sigma}^{x(n)}$. Similarly, the constraints in joint space corresponded to $\hat{\theta}$ and $\hat{\Sigma}^{\theta}$. These constraints can be mutually exclusive in the robot's workspace, i.e., the generalization in joint space does not necessarily coincide with the generalization in task space. To find a controller for the robot simultaneously satisfying several constraints, the direct computation method was used (product of Gaussian distributions). This involved computing an appropriate trade-off during the inverse kinematics process.

The reproduction procedure is described in Summary Box 4.1 and illustrated in Figure 4.4. By projecting the Gaussian distribution from task space to joint space through the Jacobian, it is here implicitly assumed that the nonlinear projection function can be approximated by the locally linear transformation $J^{\dagger}$, i.e., the local transformation is believed to remain valid for the span of data represented by the covariance matrix of the Gaussian distribution (D. Lee & Nakamura, 2007).

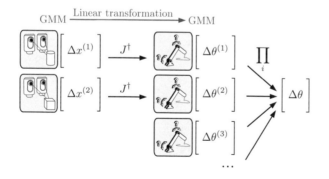

Figure 4.4 An illustration of the process used to retrieve a skill by considering constraints on various objects in task space (*first two rows*) as well as constraints in joint space (*last row*). The pseudoinverse Jacobian matrix $J^{\dagger}$ was employed to locally project the GMM representation of the constraints in task space to a corresponding representation in joint space. With the GMMs projected in joint space, an optimal solution could then be estimated through GMR by multiplying the resulting distributions using product and regression properties of Gaussian distributions.

Summary Box 4.1 The reproduction of a skill when N objects with initial positions $\{o^{(n)}\}_{n=1}^{N}$ are detected in the robot's workspace.

Offline processing and initialization

- Initialization with starting posture $\theta_0 = \hat{\theta}_0$ and end-effector location $x_0 = K(\hat{\theta}_0)$, where K is the direct kinematics function.

Loop for $t = \Delta t \to T$

    Loop for $n = 1 \to N$

- Compute the expected velocities (approximated through Euler numerical differentiation) and associated covariance matrices for the constraints relative to object n

$$
\dot{\theta}_t^{(n)} = J^{\dagger}(\theta_{t-\Delta t})\dot{x}_t^{(n)}, \quad \text{where} \quad \dot{x}_t^{(n)} = \frac{1}{\Delta t}\left[(o^{(n)} + \hat{x}_t^{(n)}) - x_{t-\Delta t}\right],
$$

$$
\Sigma_t^{(n)} = J^{\dagger}(\theta_{t-\Delta t})\,\hat{\Sigma}_t^{x(n)}\left(J^{\dagger}(\theta_{t-\Delta t})\right)^{\top}.
$$

End loop n

- Compute the expected velocity and associated covariance matrix in joint space

$$
\dot{\theta}_t^{(N+1)} = \frac{1}{\Delta t}\left[\hat{\theta}_t - \theta_{t-\Delta t}\right], \qquad \Sigma_t^{(N+1)} = \hat{\Sigma}_t^{\theta}.
$$

- Compute the new posture (and associated covariance matrix) by evaluating the product $\prod_{n=1}^{N+1} \mathcal{N}(\dot{\theta}_t^{(n)}, \Sigma_t^{(n)})$, which represents the joint probability of the considered constraints

$$
\theta_t = \theta_{t-\Delta t} + \Delta t \left[\left(\sum_{n=1}^{N+1}(\Sigma_t^{(n)})^{-1}\right)^{-1}\left(\sum_{n=1}^{N+1}(\Sigma_t^{(n)})^{-1}\dot{\theta}_t^{(n)}\right)\right],
$$

$$
\Sigma_t = \left(\sum_{n=1}^{N+1}(\Sigma_t^{(n)})^{-1}\right)^{-1}. \tag{4.15}
$$

- The new position of the end-effector is thus defined by $x_t = K(\theta_t)$.

End loop t

Consequently, a trade-off is computed in (4.15) based on the variabilities observed during the demonstrations to determine the respective relevance of the constraints in joint space and in task space. If one wishes to use a controller satisfying the constraints only in joint space, (4.15) can be replaced by $\theta_t = \theta_{t-\Delta t} + \Delta t\, \dot{\theta}_t^{(N+1)}$. Similarly, if one desires to employ a controller satisfying the constraints in task space for a specific object n, (4.15) can be replaced by $\theta_t = \theta_{t-\Delta t} + \Delta t \, \dot{\theta}_t^{(n)}$.

4.2.2   Experimental results

Figure 4.5(a) shows the five demonstrations for the two tasks, and Figures 4.6, 4.7 and 4.8 present their extracted constraints. Figure 4.5(b) provides a reproduction attempt for a new situation (with other initial positions of the objects), during which the essential features of the skill are reproduced. Figures 4.9 and 4.10 demonstrate how the constraints in joint space and task space influence the reproduction of the skill, where the final controller smoothly reproduces the essential features by adapting the extracted constraints to the new situation. At the beginning of the first task (Fig. 4.9), the actions directed toward the glass

(a) Five demonstration starting from   (b) Reproduction with a new initial
       different initial positions   situation

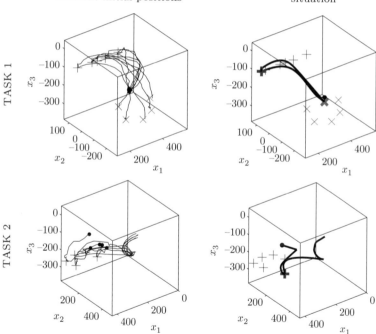

Figure 4.5 (a) Five demonstrations for the two tasks in 3D Cartesian space. For the first task, the initial positions of the glass placed on the shelf are represented with '+' signs, and the initial positions of the coaster on the table are represented with 'x' signs. For the second task, the initial positions of the glass (covering the coaster) are represented with '+' signs. (b) The reproduction of the skill for new situations (bold '+' and 'x' signs), by combining constraints in joint space and in task space. The Cartesian trajectories are represented in the robot's frame of reference (see Fig. 4.3), where the dots indicate the beginning of the motions.

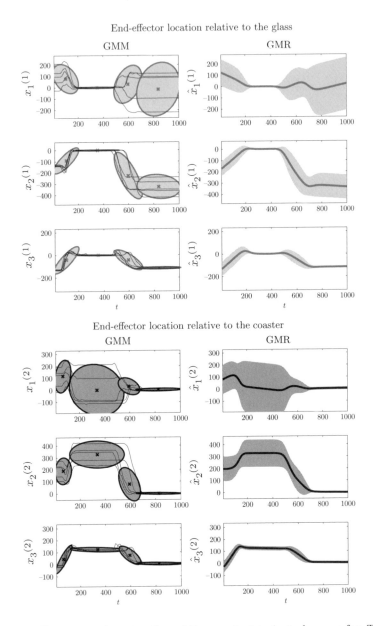

Figure 4.6 The automatic extraction of the constraints in task space for *TASK 1* (the corresponding frames of reference are depicted in Fig. 4.3), where the two sets of graphs represent the constraints relative to the various objects. GMMs with 4 Gaussian components are employed to encode the trajectories (selected by a *Bayesian Information Criterion* (BIC) for each representation). The trajectories relative to the glass appear to be highly constrained between time steps 200 and 500 when reaching for the glass. The trajectories relative to the coaster are highly constrained at the end of the motion when placing the glass on it.

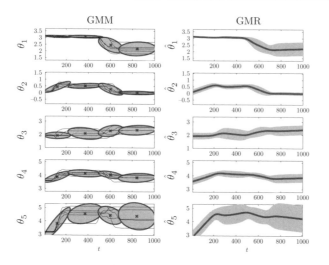

Figure 4.7 An automatic extraction of the angular constraints for *TASK 1* (the corresponding joint angles are depicted in Fig. 4.3), where GMMs with 4 Gaussian components are utilized to encode the trajectories (selected by BIC for each representation).

were of utmost importance. Subsequently, the ones directed toward the coaster predominated. It could be seen that the controller determined by the system smoothly switched from the generalized movement directed toward the glass (see e.g., x_1 at time steps 200-500) to that directed toward the coaster (see e.g., x_1 at time steps 700-1000). For the second task (Fig. 4.10), the trajectories relative to the glass were at first highly important (reaching for the glass in Cartesian space), and then gave way to a controller satisfying constraints in joint space (emptying the glass by tilting it). We can observe that the controller smoothly switched from one for which constraints in task space were important (see e.g., θ_5 at time steps 200-400) to a controller where constraints in joint space were of importance (see e.g., θ_5 at time steps 600-1000).

The location of the end-effector with respect to the objects and the non-linear combination of the joint angles used to retrieve the position of the end-effectors (direct kinematics) were hard-coded instead of being autonomously extracted by the system. This created additional redundant information in the dataset. Since manipulation tasks were considered, this information was manually included in our system by reason of its importance (i.e., the position of the end-effector in task space) being meaningful for the skill considered and remaining general for a broad range of tasks. This way, the number of examples required to extract the important aspects of the task could be reduced, thereby alleviating the need for learning the direct kinematics of the robot.[4]

---

[4] Note that the initial choice of the adequate task-dependent variables may still not be trivial for more complex paradigms.

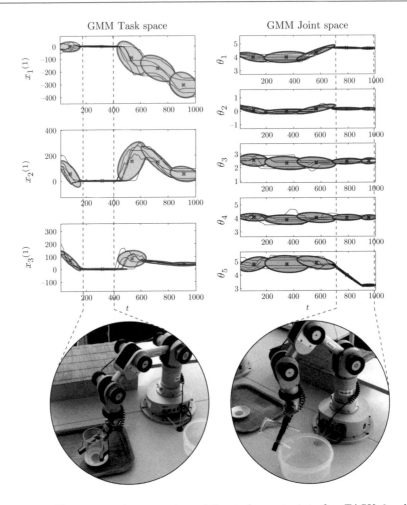

Figure 4.8 The automatic extraction of the task constraints for *TASK 2*, where GMMs with 5 Gaussians are selected by BIC to efficiently encode the skill. We see that the trajectories relative to the glass are highly constrained between time steps 200 and 400 (when reaching for the glass). Subsequently, the trajectories in joint space become more constrained. This occurs at the end of the motion, when emptying the glass in the basin by using a specific gesture. The snapshots below the graphs illustrate a reproduction attempt by automatically selecting a controller that smoothly reproduces the extracted constraints.

4.3   HANDLING OF TASK CONSTRAINTS IN LATENT SPACE – *EXPERIMENT WITH HUMANOID ROBOT*

This experiment extends the proposed approach to bimanual manipulations and demonstrates that the constraints of a skill (in joint space and task space) can

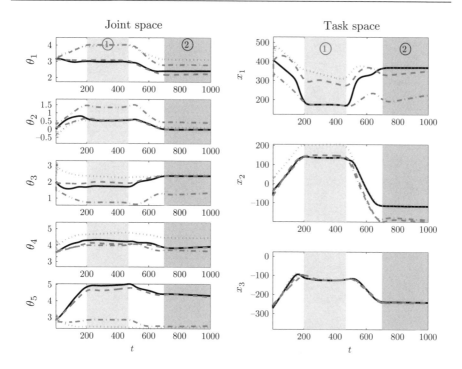

Figure 4.9 Reproduction attempts for *TASK 1* by using the extracted constraints either independently or simultaneously. The trajectories represented by *solid lines* show the final reproduction attempt by simultaneously considering the constraints in task space and in joint space. The trajectories represented by *dash-dotted lines* consider only constraints for the first object in task space. The ones displayed with *dotted lines* consider only constraints for the second object in task space. Finally, the ones represented by *dashed lines* take into account only constraints in joint space. Here, ① and ② correspond respectively to the time at which the robot grasps the glass and that at which it discards it on the coaster.

be extracted in a latent space of motion and moreover that an optimal controller can be derived from this probabilistic representation of the constraints.

Tasks with three types of constraints were here considered: (1) at the *joint angle level* (angular trajectories θ); (2) at the *hand path level* (absolute trajectories of the two hands x); and (3) at the *hands-object relationship level* (trajectories of the two hands relative to an object y).[5] For each demonstration, y is defined as a distance vector between the initial position of the object $o \in \mathbb{R}^6$ (the Cartesian position in $\mathbb{R}^3$ is defined twice for the two hands of a humanoid robot), and the position of the hands x, i.e., $y_t = x_t - o$ is computed at each time step.

---

[5] Note that the skill can, similarly, be represented as constrained by the joint angles and by actions on two objects where the first object is fixed in the environment.

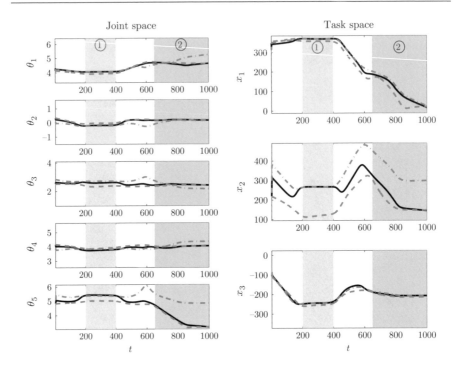

Figure 4.10 Reproduction attempts for *TASK 2* by using the extracted constraints either independently or simultaneously. The trajectories represented by *solid lines* show the final reproduction attempt. The ones shown with *dash-dotted lines* consider only constraints in task space. The ones displayed with *dashed lines* take into account only constraints in joint space. Here, ① and ② correspond respectively to the time at which the robot grasps the glass and that when it empties the glass by an appropriate tilting.

The various constraints are represented in a *latent space* of motion. By defining $\dot{x} = A^x \dot{\xi}^x$, $\dot{\theta} = A^\theta \dot{\xi}^\theta$ and $\theta = A^\theta \xi^\theta + \bar{\theta}$, the inverse kinematics problem can be rewritten as

$$\dot{x} = \tilde{J}(\theta)\,\dot{\theta}$$
$$\Leftrightarrow \quad A^x \dot{\xi}^x = \tilde{J}\left(A^\theta \xi^\theta + \bar{\theta}\right)\,A^\theta \dot{\xi}^\theta$$
$$\Leftrightarrow \quad \dot{\xi}^x = J\left(\xi^\theta\right)\,\dot{\xi}^\theta, \tag{4.16}$$
$$\text{with}\quad J\left(\xi^\theta\right) = (A^x)^\dagger \tilde{J}\left(A^\theta \xi^\theta + \bar{\theta}\right) A^\theta.$$

Here, $(A^x)^\dagger$ is the pseudoinverse of A^x, $\tilde{J}$ is the Jacobian in the original *data space*, and J is the Jacobian in the *latent space*. Similarly, the relation between $\dot{\xi}^x$ and $\dot{\xi}^y$ is defined as

$$\dot{y} = \dot{x} - \dot{o}$$
$$\Leftrightarrow \quad A^y \dot{\xi}^y = A^x \dot{\xi}^x - \dot{o}$$
$$\Leftrightarrow \quad \dot{\xi}^y = (A^y)^\dagger A^x \dot{\xi}^x - (A^y)^\dagger \dot{o}, \tag{4.17}$$

where $\dot{o}$ is the initial velocity of objects, considered to be null in this experiment. In the latent space, $\{\hat{\xi}^{\theta}, \hat{\xi}^{x}, \hat{\xi}^{y}\}$ were, respectively, set as the generalized joint angle trajectories, the generalized hand path and the generalized hands-object distance vectors extracted from the demonstrations. Let $\{\xi^{\theta}, \xi^{x}, \xi^{y}\}$ be the candidate trajectories for reproducing the motion. The metric of imitation performance (i.e., cost function for the task) $\mathcal{H}$ is given by

$$
\begin{aligned}
\mathcal{H} = {} & (\dot{\xi}^{\theta} - \hat{\dot{\xi}}^{\theta})^{\top} W^{\theta} (\dot{\xi}^{\theta} - \hat{\dot{\xi}}^{\theta}) \\
& + (\dot{\xi}^{x} - \hat{\dot{\xi}}^{x})^{\top} W^{x} (\dot{\xi}^{x} - \hat{\dot{\xi}}^{x}) \\
& + (\dot{\xi}^{y} - \hat{\dot{\xi}}^{y})^{\top} W^{y} (\dot{\xi}^{y} - \hat{\dot{\xi}}^{y}).
\end{aligned}
\tag{4.18}
$$

The problem is now reduced to finding a minimum of (4.18) when subjected to a body constraint and an environmental constraint defined respectively by (4.16) and (4.17). Since $\mathcal{H}$ is a quadratic function, the problem can be solved analytically through *Lagrange* optimization

$$
\begin{aligned}
L(\dot{\xi}^{\theta}, \dot{\xi}^{x}, \dot{\xi}^{y}) = {} & \mathcal{H} + \lambda_1^{\top}(\dot{\xi}^{x} - J\,\dot{\xi}^{\theta}) \\
& + \lambda_2^{\top}(\dot{\xi}^{y} - (A^{y})^{\dagger}A^{x}\dot{\xi}^{x} + (A^{y})^{\dagger}\dot{o}),
\end{aligned}
$$

where λ_1 and λ_2 are vectors of *Lagrange* multipliers. To solve $\nabla L = 0$, the derivations of $\frac{\partial L}{\partial \dot{\xi}^{\theta}} = 0$, $\frac{\partial L}{\partial \dot{\xi}^{x}} = 0$ and $\frac{\partial L}{\partial \dot{\xi}^{y}} = 0$ yield

$$
-2\,W^{\theta}\,(\dot{\xi}^{\theta} - \hat{\dot{\xi}}^{\theta}) - J^{\top}\lambda_1 = 0, \tag{4.19}
$$

$$
-2\,W^{x}\,(\dot{\xi}^{x} - \hat{\dot{\xi}}^{x}) + \lambda_1 - ((A^{y})^{\dagger}A^{x})^{\top}\lambda_2 = 0, \tag{4.20}
$$

$$
-2\,W^{y}\,(\dot{\xi}^{y} - \hat{\dot{\xi}}^{y}) + \lambda_2 = 0. \tag{4.21}
$$

Combining (4.20) and (4.21) provides

$$
\lambda_1 = 2\,W^{x}\,(\dot{\xi}^{x} - \hat{\dot{\xi}}^{x}) + ((A^{y})^{\dagger}A^{x})^{\top}\,2\,W^{y}\,(\dot{\xi}^{y} - \hat{\dot{\xi}}^{y}). \tag{4.22}
$$

Combining (4.22) and (4.19) yields

$$
W^{\theta}\,(\dot{\xi}^{\theta} - \hat{\dot{\xi}}^{\theta}) + J^{\top}W^{x}\,(\dot{\xi}^{x} - \hat{\dot{\xi}}^{x}) + J^{\top}\,((A^{y})^{\dagger}A^{x})^{\top}\,W^{y}\,(\dot{\xi}^{y} - \hat{\dot{\xi}}^{y}) = 0,
$$

and solving for $\dot{\xi}^{\theta}$ gives

$$
\begin{aligned}
\dot{\xi}^{\theta} = {} & \left(W^{\theta} + J^{\top}W^{x}J + ((A^{y})^{\dagger}A^{x}J)^{\top}W^{y}((A^{y})^{\dagger}A^{x}J)\right)^{-1} \\
& \cdot \left(W^{\theta}\,\hat{\dot{\xi}}^{\theta} + J^{\top}W^{x}\,\hat{\dot{\xi}}^{x} + ((A^{y})^{\dagger}A^{x}J)^{\top}W^{y}(A^{y})^{\dagger}\dot{o} + \hat{\dot{\xi}}^{y}\right).
\end{aligned}
$$

The joint angle trajectories can then be reconstructed iteratively with $\xi_t^{\theta} = \xi_{t-\Delta t}^{\theta} + \Delta t\,\dot{\xi}_t^{\theta}$ and $\theta_t = A^{\theta}\xi_t^{\theta} + \bar{\theta}$.

4.3.1 Experimental setup

Based on the process described above, three tasks were taught to a *HOAP-2* humanoid robot, of which only the 11 DOFs of the arms and torso were required in the experiments (8 DOFs for the arms, 1 DOF for the torso, and 2 binary commands to open and close the robot's hands). The remaining DOFs of the legs were set to a constant position to support the robot in an upright posture facing a table. The three tasks are illustrated in Figure 4.11, and the demonstrations were provided through kinesthetic teaching. The initial position of the objects were registered by helping the robot grasp and release them prior to the demonstration of the skill. During this *molding* process, the teacher grabbed one of the robot's arm, moved it toward the object, positioned the robot's palm around the object, pressing its fingers against it to inform the robot that an object was currently in its palm. When the force sensor located within the robot's palm retrieved a value above a given threshold, the robot briefly grasped and released the object, thereby registering its position in 3D Cartesian space (the robot remained seated during the experiment).

It was assumed, in this experiment, that the important sensory information came from: (1) the posture of the robot defined by the joint angles provided by the motor encoders actuating the upper-body part; (2) the absolute positions of the hands of the robot in Cartesian space calculated through direct kinematics from the joint angles; (3) the relative positions between the hands and the object calculated from the absolute initial position of the object and the absolute positions of the hands in Cartesian space; and (4) the open/close status of the two hands of the robot.

The first three datasets were first projected in latent spaces of motion by means of PCA, as described in Section 2.7.1. The various trajectories were then aligned temporally using the DTW approach introduced in Section 2.9.3. The continuous trajectories were encoded in GMM through a batch learning procedure, as presented in Section 2.6.1. Generalized versions of these trajectories were retrieved through GMR, as detailed in Section 2.4. The open/close status of the two hands were encoded in BMM, as presented in Section 2.11.1.

The robot was provided with 4 to 7 demonstrations by an expert user where each demonstration was rescaled to $T = 100$ datapoints. Once trained, the robot was required to reproduce each task under various constraints by placing the object at several locations within its workspace. This procedure aimed at demonstrating the robustness of the system when the constraints were transposed to different locations within the robot's workspace.

4.3.2 Experimental results

Figures 4.12 and 4.13 provide an example for the *Bucket Task* to respectively select the dimensionality of the latent space and the number of GMM components used to encode the projected data. The results for the other tasks are summarized in Table 4.1. It can be seen that the dimensionalities of the latent space for the *"Chess Task"* and *"Bucket Task"* were higher than for the *"Sugar Task"*, probably because both arms were used in these cases (for the *"Chess*

"Chess Task"

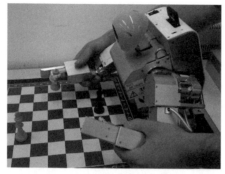

"Bucket Task"

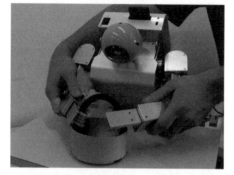

"Sugar Task"

Figure 4.11 An illustration of the kinesthetic teaching process for the 3 tasks considered in this experiment. *"Chess Task"*: Grabbing and moving a chess piece two squares forward. *"Bucket Task"*: Grabbing and bringing a bucket to a specific position. *"Sugar Task"*: Grabbing a lump of sugar and bringing it to the mouth, using either the right or left hand.

Task", the left hand was used only to maintain stability while the robot leaned over the table). For the *"Sugar Task"*, one hand was usually motionless while the other performed the task. The number of Gaussians extracted by the Bayesian Information Criterion (BIC) ranged from 4 to 7. One should note

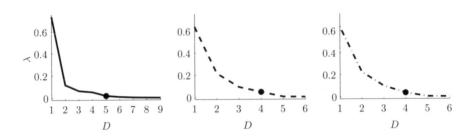

Figure 4.12 An estimation of the dimensionality required when reducing the number of variables for the *"Bucket Task"* through Principal Component Analysis (*solid line* for the joint angles, *dashed line* for the hand paths and *dash-dotted line* for the hands-object relationship). The dots correspond to the number of dimensions representing at least 98% of the data variance.

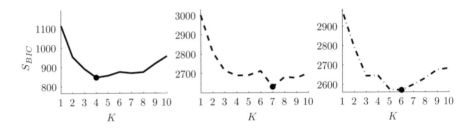

Figure 4.13 An estimation of the numbers of Gaussians in the GMMs required to model the trajectories in latent space for the *"Bucket Task"* (*solid line* for the joint angles, *dashed line* for the hand paths and *dash-dotted line* for the hands-object relationship). Adding more Gaussians increases the log-likelihood but also increases the number of parameters. The dots represent the minima determined by the Bayesian Information Criterion used to select an optimal number of parameters.

that the total number of parameters depended quadratically on the dimensionality of the latent space and linearly on the number of Gaussians, i.e., the total number of parameters used to encode the data was $n_{PCA} = (D-1)(d-1)$ and $n_{GMM} = (K-1) + K\left(D + \frac{1}{2}D(D+1)\right)$. A single component was found by BMM to efficiently encode the binary signals h of the hands since they were closed simultaneously for each considered task.

Figures 4.15 and 4.16 present the reproduced trajectories of the *"Chess Task"* for various initial positions of the chess piece (see also the legend of Fig. 4.14). Given that the right shoulder was located at $(100, -35)$, we see on the map of Figure 4.16 that the best location to reproduce the motion involved initially placing the chess piece in front of the right arm (see graph ①). In graphs ② and ③, the chess piece was initially placed at various positions unobserved during the demonstrations. Figures 4.17 and 4.18 show the reproduced trajectories of the *"Bucket Task"* for a variety of initial position of the bucket. The optimal trajectory satisfying the learned constraints was globally found to follow the demonstrated hand paths, still using the demonstrated

Table 4.1 The number of parameters automatically extracted by the system.

		Data space			Latent space	
		n	T	d	D	K
"Chess Task"	θ	7	100	9	4	5
obj1: chess board (fixed)	x	7	100	6	4	5
obj2: chess piece	y	7	100	6	4	4
	h	7	100	2	(2)	1
"Bucket Task"	θ	7	100	9	5	4
obj1: table (fixed)	x	7	100	6	4	7
obj2: bucket	y	7	100	6	4	6
	h	7	100	2	(2)	1
"Sugar Task" - left	θ	4	100	9	2	5
obj1: mouth (fixed)	x	4	100	6	3	5
obj2: piece of sugar	y	4	100	6	3	6
	h	4	100	2	(2)	1
"Sugar Task" - right	θ	4	100	9	2	6
obj1: mouth (fixed)	x	4	100	6	3	6
obj2: piece of sugar	y	4	100	6	3	6
	h	4	100	2	(2)	1

```
——— Hand path reproduced by the robot
- - - Generalized hand path x̂
······· Hand path reconstructed from the generalized joint angle trajectory θ̂
  O   Initial position of the object o
  X   Initial position(s) of the hand(s)
```

Figure 4.14 Legends for the graphs presented in Figures 4.15-4.20.

object-hands trajectories when approaching the bucket.[6] Figures 4.19 and 4.20 demonstrate the reproduced trajectories of the *"Sugar Task"* for various initial positions of the lump of sugar. In this task, the table was centered in front of the robot who was taught two gestures: (1) how to, with its right hand, grasp the lump of sugar located to the far right of the table; (2) how to, with its left hand, grasp a lump of sugar located to the far left of the table. A model of the skill was created for both the left and right arm. Depending on the situation, an optimal controller for the reproduction was then selected by evaluating each metric respectively. We see in the bottom-right inset of Figure 4.20 that the limit between the left/right controller was clearly situated at $x_1 = 0$, i.e., on the symmetry axis of the robot.

By projecting the original data onto a latent space of motion and encoding the resulting data in GMM, the system was able to generalize the skill over different situations (varying initial positions of the object). The generalization capabilities depended directly on the variations in the demonstrations provided

---

[6] An example of failure for this task will be presented further in Section 8.2.3.

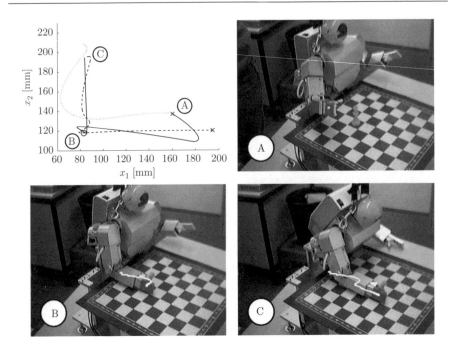

Figure 4.15 The reproduction of the *"Chess Task"* with an initial position of the object close to the generalized trajectories. The three snapshots represent the motion at various time steps Ⓐ, Ⓑ and Ⓒ, also represented on the graph. Moreover, the snapshots show the Cartesian trajectory of the hand tracked by the stereoscopic vision system (see legend in Fig. 4.14).

to the robot as well as on the dimensionality of the latent space obtained by the system. For example, if the system detected that the hands were constrained to moving an object in a plane (2-dimensional hand path latent space), the system was further constrained to the reproduction of only planar motions. For small changes in the initial position of the object between the demonstrations and the reproduction, the robot was capable of correctly adapting its gestures and reproducing the important characteristics of each task, namely: (1) grasping and moving the chess piece with a specific relative path; (2) grasping the bucket with two hands and moving it to a specific location; (3) grasping the lump of sugar and bringing it to its mouth with either its left or right arm. These essential features could be reproduced even if none of these high-level goals were explicitly represented in the robot's control system.

In conclusion, the essential features of the three tasks were successfully extracted using a continuous representation of the skill, and it was demonstrated that an optimal controller for the reproduction could be computed by deriving a metric of imitation, which, for each task, was able to detect a solution with the best possible match to the various trajectories demonstrated by the user. Depending on the task, these variables had different importance, which could be estimated by observing a human expert demonstrating the skill several

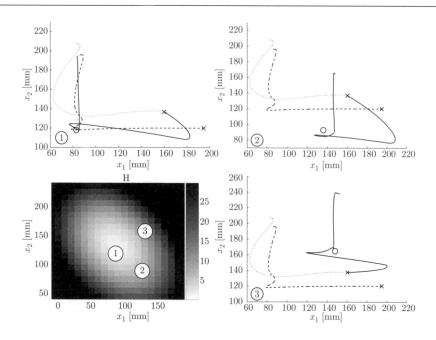

Figure 4.16 *Bottom left:* A map depicting values of the cost function $\mathcal{H}$ for the *"Chess Task"* with respect to the initial position of the chess piece on the board. The graphs ①, ② and ③ illustrate the reproduction attempts for the corresponding locations on the map (see legend in Fig. 4.14).

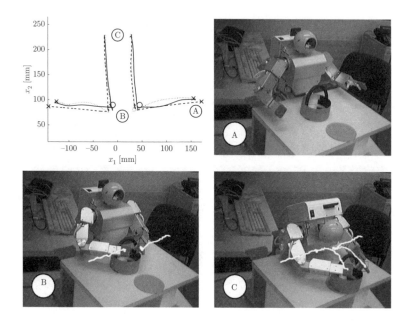

Figure 4.17 The reproduction of the *"Bucket Task"* with an initial position of the object close to the generalized trajectories. The three snapshots represent the motion at various time steps Ⓐ, Ⓑ and Ⓒ, also displayed on the graph.

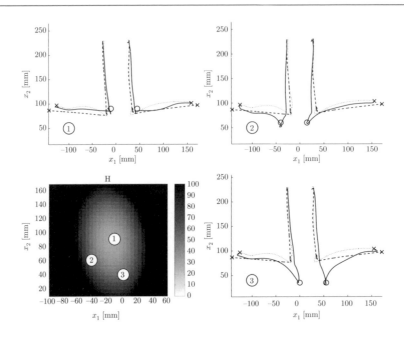

Figure 4.18 *Bottom left:* A map depicting values of the cost function $\mathcal{H}$ for the *"Bucket Task"*, with respect to the initial position of the bucket on the table. The graphs ①, ② and ③ show the reproduction attempts for the corresponding locations on the map.

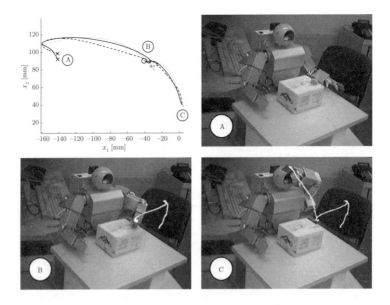

Figure 4.19 The reproduction of the *"Sugar Task"* with an initial position of the object close to the generalized trajectories. The three snapshots illustrate the motion at various time steps Ⓐ, Ⓑ and Ⓒ, also represented on the graph.

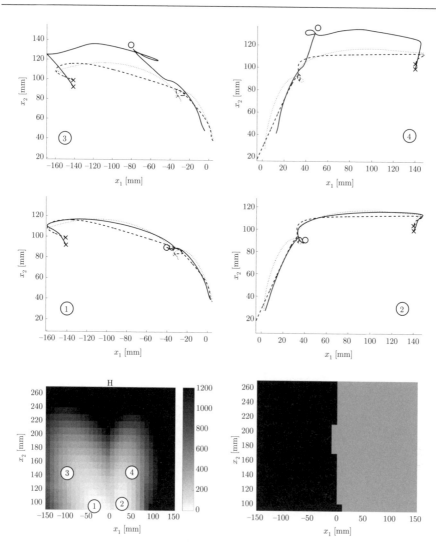

Figure 4.20 *Bottom left:* A map depicting values of the cost function $\mathcal{H}$ for the *"Sugar Task"*, with respect to the initial position of the lump of sugar on the box. The graphs ①, ②, ③ and ④ show the reproduction attempts for the corresponding locations on the map (see legend in Fig. 4.14). Two areas are visible for the left and right arm. *Bottom right:* The selection of a left/right arm controller depending on the value of $\mathcal{H}$ (the black area corresponds to $\mathcal{H}_{\text{left}} < \mathcal{H}_{\text{right}}$).

times in varying situations. The number of examples required for an efficient reproduction of the task depends on the teaching efficiency of the user. Indeed, an expert teacher may produce demonstrations that explore as much as possible the variations allowed by the task, while a naive user may demonstrate a task several times without fully exploiting the alternatives allowed. This issue is discussed in detail in Chapter 6.

CHAPTER 5

EXTENSION TO DYNAMICAL SYSTEM AND HANDLING OF PERTURBATIONS

In previous chapters, we demonstrated that the GMM/GMR representation could benefit from the explicit time-precedence of the motion segments to ensure a precise reproduction of the task. We described applications where GMR was used to retrieve a smooth generalized version of a set of trajectories, representing either joint angle variables, Cartesian coordinates, or both. For trajectories in Cartesian space, the joint distribution $\mathcal{P}(t, x)$ was represented in a GMM, where t constituted time components and x corresponded to position components. The conditional probability $\mathcal{P}(x|t)$ was then estimated through GMR along the whole range of time steps $t \in [0, T]$. This was used to retrieve another set of Gaussian distributions representing the expected spacial values for each time step. A smooth trajectory could thus be constructed by aggregating the centers of the Gaussian distributions. However, it was also demonstrated that this explicit time-dependence required an appropriate segmentation of the various demonstrations or/and the use of methods for realigning and scaling the trajectories to handle perturbation.

As an alternative, recent approaches have considered modeling the intrinsic dynamics of motion (Ijspeert, Nakanishi, & Schaal, 2002b; Grimes, Chalodhorn, & Rao, 2006). Such approaches are advantageous in that the system is time-independent and can be modulated to produce trajectories with similar dynamics in areas of the workspace not covered during training. However, such techniques either assumed a basic form for the dynamical system to be learned or explored the adaptivity of the system locally around its stable points.

The present chapter suggests exploiting the strength of Gaussian Mixture Regression in order to represent the dynamics of the motions.[1] The regression approach is thus extended to a dynamical system by encoding $\mathcal{P}(x, \dot{x})$ in a GMM and by iteratively retrieving $\mathcal{P}(\dot{x}|x)$ through GMR.

---

[1] Note that we here employ a loose definition of the term *dynamics* to contrast the joint use of position and velocity to the use of only position.

5.1 PROPOSED DYNAMICAL SYSTEM

Multiple examples of a skill are demonstrated to the robot in slightly different situations, where position x and velocities $\dot{x}$ are collected during the demonstrations. The dataset is composed of a series of datapoints $\xi = [x, \dot{x}]$, where the joint distribution $\mathcal{P}(x, \dot{x})$ is encoded in a GMM of K Gaussians, locally representing the correlations between the variables. Input and output components can be represented separately through the notation

$$\mu_i = \begin{bmatrix} \mu_i^x \\ \mu_i^{\dot{x}} \end{bmatrix} \quad \text{and} \quad \Sigma_i = \begin{bmatrix} \Sigma_i^x & \Sigma_i^{x\dot{x}} \\ \Sigma_i^{\dot{x}x} & \Sigma_i^{\dot{x}} \end{bmatrix},$$

where the uppercase indices x and $\dot{x}$ refer respectively to position and velocity components.

At each iteration, for a given datapoint ξ, corresponding to a current position x and velocity $\dot{x}$, a weight factor h_i representing the influence of the i-th Gaussian distribution is defined as

$$h_i(\xi) = \frac{\mathcal{N}(\xi; \mu_i, \Sigma_i)}{\sum_{k=1}^{K} \mathcal{N}(\xi; \mu_k, \Sigma_k)} \quad . \tag{5.1}$$

A desired position $\hat{x}$ and velocity $\hat{\dot{x}}$ to attain are estimated through GMR as

$$\hat{x} = \sum_{i=1}^{K} h_i(\xi) \left(\mu_i^x + \Sigma_i^{x\dot{x}} (\Sigma_i^{\dot{x}})^{-1} (\dot{x} - \mu_i^{\dot{x}}) \right),$$

$$\hat{\dot{x}} = \sum_{i=1}^{K} h_i(\xi) \left(\mu_i^{\dot{x}} + \Sigma_i^{\dot{x}x} (\Sigma_i^{x})^{-1} (x - \mu_i^x) \right).$$

From the current position and velocity of the system, an impedance controller is computed (similarly as a mass-spring-damper system) to reach the desired velocity $\hat{\dot{x}}$ and the desired position $\hat{x}$. This is demonstrated in Figure 5.1.[2] The acceleration command is determined by

$$\ddot{x} = \overbrace{(\hat{\dot{x}} - \dot{x})\kappa^{\mathcal{V}}}^{\ddot{x}^{\mathcal{V}}} + \overbrace{(\hat{x} - x)\kappa^{\mathcal{P}}}^{\ddot{x}^{\mathcal{P}}}. \tag{5.2}$$

The effect of the impedance controller parameters on the dynamic behavior is illustrated in Figure 5.2. Subsequently, velocity and position are updated at each iteration through Euler numerical integration[3]

$$\dot{x}_t = \dot{x}_{t-\Delta t} + \Delta t \, \ddot{x},$$

$$x_t = x_{t-\Delta t} + \Delta t \, \dot{x}_t. \tag{5.3}$$

---

[2] Note that this controller is also similar the second-order differential equation defined by a *Vector Integration To Endpoint* (VITE) system (Bullock & Grossberg, 1988).

[3] Note that it is also possible to employ other numerical methods for ordinary differential equations can also be used (Butcher, 2003).

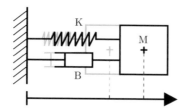

Figure 5.1 Virtual mass-spring-damper system $M\ddot{x} + B(\dot{x} - \dot{x}_0) + K(x - x_0) = 0$ with mass M, viscous damper of coefficient B and spring constant K used to model an impedance controller of the form $\ddot{x} = (\dot{\hat{x}} - \dot{x})\kappa^{\mathcal{V}} + (\hat{x} - x)\kappa^{\mathcal{P}}$.

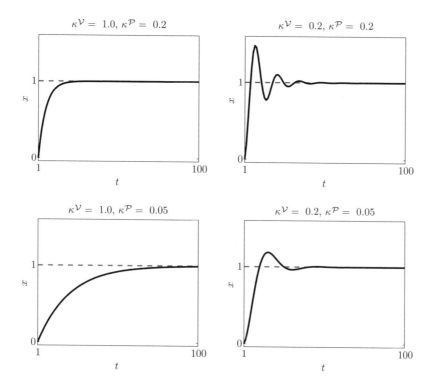

Figure 5.2 The behavior of the mass-spring-damper system for various sets of $\kappa^{\mathcal{V}}$ and $\kappa^{\mathcal{P}}$ parameters. The initial state of the system is defined by $x = 0$ and $\dot{x} = 0$. The position target and velocity target are defined by $\hat{x} = 1$ and $\dot{\hat{x}} = 0$. Depending on the parameters, the system presents several dynamical behaviors in terms of reactivity, smoothness and overshoot.

In (5.2), $\ddot{x}^{\mathcal{V}}$ allows to follow the demonstrated dynamics and $\ddot{x}^{\mathcal{P}}$ prevents the robot from moving far away from an unlearned situation and to come back to an already encountered context if a perturbation occurs. By using both terms concurrently, the robot follows the learned non-linear dynamics while coming back to a known motion if it deviates from the demonstrated motion and arrives

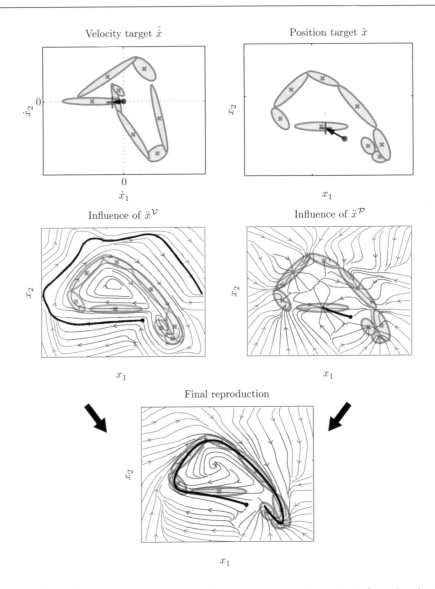

Figure 5.3 The dynamical system used to reproduce a demonstrated motion in a robust manner.

in an undiscovered portion of the workspace. An illustration of the complete process is presented in Figure 5.3. The first row portrays how to determine components that reach a desired velocity $\hat{\dot{x}}$ and desired position $\hat{x}$ through a second-order dynamical system. The vector fields in the second row display the influence of the two commands when used separately. On the one hand, $\ddot{x}^{\mathcal{V}}$ follows the learned motion but tends to move away from the demonstrations after a few iterations or when starting from an unexplored position. On the other hand, $\ddot{x}^{\mathcal{P}}$ moves toward the closest point of the generalized trajectory. An

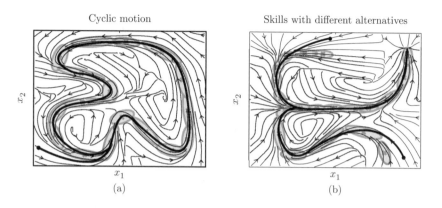

Figure 5.4 Illustrations of the proposed dynamical system in situations presenting cyclic motions (a) and where several alternatives of motions are demonstrated to the robot (b). The points show the initial positions.

example of reproduction is also depicted with a thick line. The vector field in the last row shows the reproduction behavior when simultaneously considering both commands. The final controller follows the demonstrated dynamics and prevents the robot from moving away from an unlearned situation by coming back to an already encountered position if it deviates from the demonstrated motion. Figure 5.4 illustrates the use of the system with cyclic motions and skills presenting several alternatives (i.e., where the robot reproduces the skill differently depending on the initial situation from which the motion starts).

5.1.1 Extension to motions containing pauses and loops

The system presented above suffers from two drawbacks. First, if the demonstrated motion displays a null velocity between the start and the end of the motion, the robot may stop at this point during reproduction due to no sequential information being taken into account when computing the influence of the Gaussian distributions in (5.1). Similarly, if the motion presents a loop, the system may have problems correctly handling the redundant region as a result of it not having the sequential information required to unambiguously describe the dynamics. Indeed, the weight computed in (5.1) depends on the current position and velocity but does not keep track of the previous positions and velocities retrieved by the system.

In order to add the sequential information required to properly deal with these types of signals, the proposed approach is extended to the use of a *Hidden Markov Model* (HMM) to replace the weight computed in (5.1). The sequence of Gaussian distributions is encoded in an HMM where each state is represented by one Gaussian distribution. The parameters of the model are $\{\Pi, a, \mu, \Sigma\}$, where Π_i is the initial probability of being in state i, a_{ij} is the probability of transiting from state i to state j, and μ_i and Σ_i represent the center and covariance matrix of the i-th Gaussian distribution of the GMM with K Gaussians (see Sect. 2.3). If the GMM has already been learned, only the initial state distribution Π_i and transition probabilities a_{ij} of the HMM need to be estimated (see Sect. 2.6.2). A simpler alternative consists in setting the initial state distribution Π to equally distributed weights (rendering it possible to start from any point in space during reproduction) and setting the transition probabilities a_{ij} to a left-right model with pre-determined weights for the self-transition probability and the transition to the next Gaussian distribution.

By using the *forward algorithm* presented in Section 2.3.1, the weight described in Eq. (5.1) is redefined as a weight based on both the spatial and the sequential aspects

$$h_i(\xi_t) = \frac{\alpha_{i,t}}{\sum_{k=1}^{K} \alpha_{k,t}}, \quad \text{with } \alpha_{i,t} = \left(\sum_{k=1}^{K} \alpha_{k,t-1} \, a_{ki} \right) \mathcal{N}(\xi_t; \, \mu_i, \Sigma_i), \quad (5.4)$$

where $\alpha_{i,t}$ is the *forward* variable (defined recursively) representing the probability of partially observing a sequence $\{\xi_1, \xi_2, \dots, \xi_t\}$ of length t and of being in state i at time t.

The resulting system then probabilistically considers the sequential nature of the learned motion. Figure 5.5 illustrates the behavior of the system with and without this HMM representation when taking into account loops or pauses in the motion.

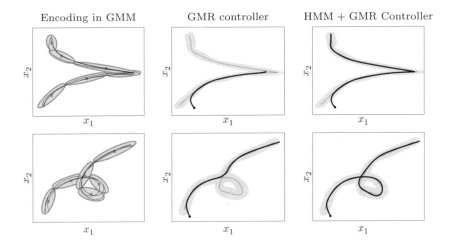

Figure 5.5 Examples of motions with pauses and loops. The first row presents a motion that starts from the bottom of the graph, stops in the middle, and then moves upwards. The second row presents a motion with a loop starting from the bottom of the graph. The first column shows the model of the motion in task space. The second column displays the behavior of the system when computing the weights with only spatial information, as proposed in Eq. (5.1). The last column demonstrates the behavior of the system when sequential and spatial information are taken into account for the computation of the weights, as proposed in Eq. (5.4).

5.2   INFLUENCE OF THE DYNAMICAL SYSTEM PARAMETERS

To systematically analyze the influences of the various parameters, several sets of trajectories are created by randomly generating a series of N keypoints X of D dimensions, where each variable $\{X_i\}_{i=1}^{D}$ has a uniform random distribution $\mathcal{U}(0,1)$. A mass-spring-damper controller with parameters $\kappa^V = 1.0$ and $\kappa^P = 0.03$ is then used to obtain trajectories by starting from the first keypoint and defining the second keypoint as the target. Every 25 steps, the target is switched to the next keypoint. For the last keypoint, 25 additional steps are used. To simulate motion variability, each dataset consists of 3 trajectories produced in this manner by slightly varying the positions of the keypoints with a Gaussian noise $\mathcal{N}(0, 0.1)$. A 2-dimensional example with $N = 4$ keypoints is presented in Figure 5.6.

Two metrics are then used to evaluate the reproduction attempts characterized by positions x' and velocities $\dot{x}'$. a root mean square error on position and velocity is computed along the motion as well as its end with

$$\mathcal{E}_1 = \frac{1}{T}\sum_{t=1}^{T} \left|\left| x'_t - \frac{1}{3}\sum_{i=1}^{3} x_{i,t} \right|\right|, \qquad \mathcal{E}_2 = \frac{1}{T}\sum_{t=1}^{T} \left|\left| \dot{x}'_t - \frac{1}{3}\sum_{i=1}^{3} \dot{x}_{i,t} \right|\right|.$$

The influence of the dynamical system parameters are presented in Figure 5.7. It can be observed that values of $\kappa^V = 1.0$ and $\kappa^P = 0.03$ provide good results for both $\mathcal{E}_1$ and $\mathcal{E}_2$. For further experiments, these parameters will be used as a compromise between following the dynamics and returning to the learned motion. Figure 5.8 shows the influences of the dimensionality D, the number of keypoints N (i.e., the complexity of the curve) and the number of states K in the GMM. The first graph demonstrates that the dimensionality D has a very low and linear impact on the performance of the system (for a dimensionality lower than 25). Increasing the complexity of the curve by maintaining the same number of Gaussians in the GMM shows a degradation of the performance (*second graph*). By increasing the number of states K, the error reduces until a limit is attained, at which the addition of Gaussians has very little influence (*third graph*). If the number of Gaussians increases with the number of keypoints (by considering that two Gaussians are required per keypoint), the error remains low (*last graph*).

5.3   EXPERIMENTAL SETUP

Two experiments were conducted with a *HOAP-3* humanoid robot from *Fujitsu* of which the 10 DOFs of the upper torso were used in the experiment (4 DOFs per arm and 2 DOFs for the head). A *kinesthetic teaching* process was employed to record motion, through motor encoders of the robot registering information while the teacher moved the robot's arms. The selected motors were set to passive mode, which allowed the user to freely move the corresponding degrees

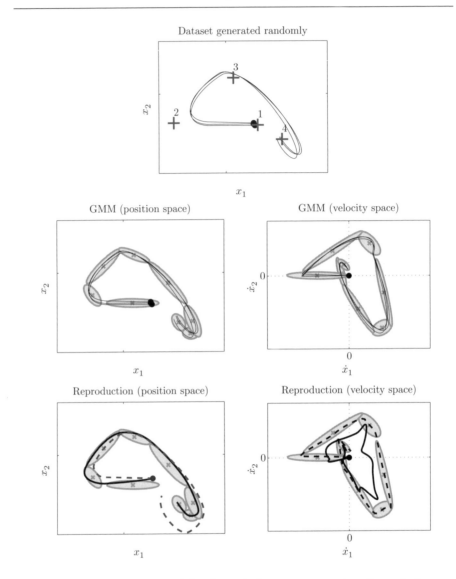

Figure 5.6 The optimization of the dynamical system parameters through the generation of a set of trajectories with randomly generated keypoints. *First row:* Data randomly generated through a series of $N = 4$ keypoints. *Second row:* Learned GMM represented in position and velocity space. *Last row:* Two reproduction attempts by varying the parameters of the impedance controller to show their influences in the position and velocity spaces. The motion represented with a solid line uses high gain for $\kappa^{\mathcal{P}}$ (i.e., it follows the demonstrated path closely) while the dashed line corresponds to a motion using high gain for $\kappa^{\mathcal{V}}$ (i.e., it follows the demonstrated velocities closely).

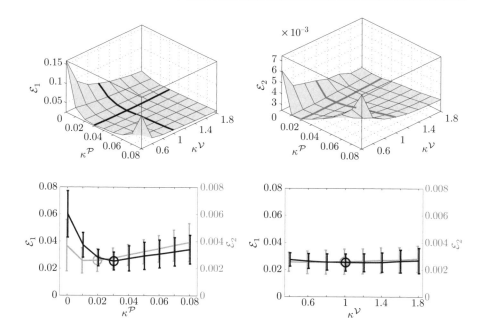

Figure 5.7 The influence of the impedance controller parameters on the reproduction error $\mathcal{E}_1$ (*black line*) and $\mathcal{E}_2$ (*gray line*) for different values of $\kappa^{\mathcal{V}}$ and $\kappa^{\mathcal{P}}$ (*top*), and around the minimal error values given by $\kappa^{\mathcal{V}} = 1.0$ and $\kappa^{\mathcal{P}} = 0.03$ (*bottom*), with a dataset generated with $D = 3$ (dimensionality), $N = 4$ (number of keypoints) and $K = 8$ (number of Gaussians). In the bottom graphs, the circled results depict the minimal error values.

of freedom while the robot executed the task. The kinematics of each joint motion were recorded at a rate of 1000 Hz, and the registered trajectories were resampled to a fixed number of 100 datapoints.

The first experiment involved feeding a *Robota* robotic doll (Billard, 2003) and consisted in learning the proper gesture required to move a spoon, bring it to *Robota*'s mouth and then move the spoon away from the mouth. The *HOAP-3* robot holds *Robota* by the shoulder during the learning and reproduction processes. This allows the robot to precisely register the position and orientation of Robota's mouth. Robota is endowed with a motor and potentiometer at the level of the head. During the demonstration phase, the potentiometer is used to record the mouth position and orientation through direct kinematics. During reproduction, the experimenter manually moves HOAP-3's left arm and turns Robota's head. The robot then adapts to this change in an online manner. The aim of this experiment was to demonstrate the robustness of the system to various initial conditions and to perturbations occurring in the course of reproduction.

The second experiment was similar to the first, except that it involved two different landmarks. The setup of this experiment is presented in Figure 5.9.

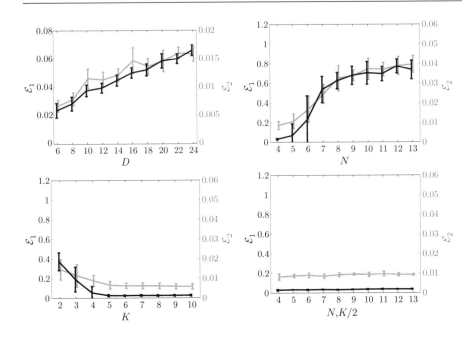

Figure 5.8 The influence of the dimension, number of states and number of keypoints on the reproduction errors $\mathcal{E}_1$ (*black line*) and $\mathcal{E}_2$ (*gray line*) for initial values $N = 4$ (number of keypoints), $D = 6$ (dimensionality) and $K = 8$ (number of Gaussians). Note that the scaling for the first graph (influence of dimensionality) has been reduced by an order of magnitude.

HOAP-3 first brings the spoon towards its own mouth and then towards *Robota*'s mouth. The aim of this experiment was to demonstrate that multiple objects can be considered in the learning framework.

Three demonstrations were provided for each experiment by moving the initial positions of the landmarks from one demonstration to the other. The experimenter notified the start and the end of the recording to the robot.

5.3.1   Illustration of the problem

For didactic purposes, the description of the above-mentioned problem is sketched in Figure 5.9. It serves as a case study to drive the argument of the chapter, however this particular implementation does not restrict the generality of the method presented here. Let us, thus, consider a sequential task in which a robot first feeds itself and then feeds another robot. Such a task can be decomposed into a first motion, consisting in moving a spoon towards the robot's own "mouth", followed by the motion of moving the spoon from the robot's mouth to that of the second robot. Since here our interest lies in learning tasks as generically as possible, avoiding to provide the robot with *a priori* knowledge on any of the constraints of this skill, we consider that only the

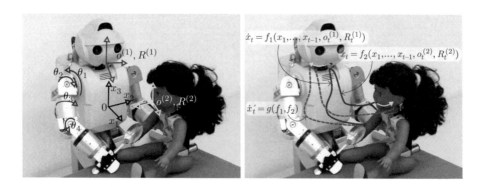

Figure 5.9 The system setup and problem statement.

position $o^{(1)}$ and orientation $R^{(1)}$ of the robot's mouth, and the position $o^{(2)}$
and orientation $R^{(2)}$ of the second robot's mouth are known. Through multiple
observations, a control law of the form $\dot{x} = f_1(x)$ is learnt to first reach for its
mouth (by using a specific gesture) and then move away in the direction of the
other robot's mouth through a controller $\dot{x} = f_2(x)$. When reproducing the
skill in a new situation (represented by arrows in the snapshot), the robot must
find a controller $\dot{x} = g(f_1(x), f_2(x))$ that optimally satisfies the two constraints
associated with these two landmarks. It must be able to start from different
postures and to deal with dynamic perturbations while reproducing the skill.
In addition, it must find an appropriate set of joint angle trajectories to fulfill
the task requirements in Cartesian space.

5.3.2   Handling of multiple landmarks

The dynamical system presented above can be extended to a controller simultaneously combining multiple constraints, i.e., where the position of the end-
effector is described with respect to a set of landmarks (that are pre-defined
by the experimenter) or to a series of objects detected by the robot in its environment. An illustration of the learning and reproduction process is presented
in Figure 5.10.

In the demonstration phase, the position x of the end-effector is collected
in the frame of reference of the robot's torso (fixed frame of reference). This
trajectory can be expressed in the frames of reference of the various landmarks
(moving frames of references) defined for each landmark n by position $o^{(n)}$ and
orientation matrix $R^{(n)}$

$$x^{(n)} = \left(R^{(n)}\right)^{-1} \left[x - o^{(n)}\right] , \qquad \dot{x}^{(n)} = \left(R^{(n)}\right)^{-1} \dot{x}.$$

Encoding in a *Gaussian Mixture Model* is then performed for each land-
mark on the resulting trajectories $\xi^{(n)} = [x^{(n)}, \dot{x}^{(n)}]$. During the reproduction
phase, the current position and velocity is projected at each iteration in the
frames of reference of the landmarks, and a position target $\hat{x}^{(n)}$ and velocity

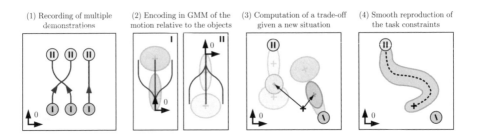

Figure 5.10 The handling of multiple objects in the probabilistic framework. (1) Three demonstrations x are provided to teach the motion of starting from landmark I and moving upward to landmark II. (2) Based on the trajectories associated to the two landmarks, a GMM with distributions $\mathcal{N}(\mu^{(n)}, \Sigma^{(n)})$ is created for each landmark n. (3) For a new position and orientation of landmark n, each distribution $\mathcal{N}(\mu^{(n)}, \Sigma^{(n)})$ is projected back to the fixed frame of reference and a command is computed at each iteration by taking into account the two constraints through a Gaussian product. (4) A smooth trajectory is retrieved by repeating the process incrementally.

target $\hat{x}^{(n)}$ (with their associated covariances $\hat{\Sigma}^{x^{(n)}}$ and $\hat{\Sigma}^{\dot{x}^{(n)}}$) are defined for each landmark, as described previously. For the current positions $o'^{(n)}$ and orientations $R'^{(n)}$ of the landmarks, $\hat{x}^{(n)}$ and $\hat{\dot{x}}^{(n)}$ are projected back to the frame of reference attached to the torso

$$\hat{x}'^{(n)} = R'^{(n)} \hat{x}^{(n)} + o'^{(n)} , \qquad \hat{\dot{x}}'^{(n)} = R'^{(n)} \hat{\dot{x}}^{(n)},$$

with associated covariance matrices converted through the linear transformation property of Gaussian distributions stated in Eq. (2.5)

$$\hat{\Sigma}'^{x^{(n)}} = R'^{(n)} \, \hat{\Sigma}^{x^{(n)}} \, \left(R'^{(n)}\right)^{\top} , \qquad \hat{\Sigma}'^{\dot{x}^{(n)}} = R'^{(n)} \, \hat{\Sigma}^{\dot{x}^{(n)}} \, \left(R'^{(n)}\right)^{\top}.$$

The final position target $\hat{x}$ and velocity target $\hat{\dot{x}}$ can be computed through the Gaussian products $\prod_{n=1}^{N} \mathcal{N}(\hat{x}'^{(n)}, \hat{\Sigma}'^{x^{(n)}})$ and $\prod_{n=1}^{N} \mathcal{N}(\hat{\dot{x}}'^{(n)}, \hat{\Sigma}'^{\dot{x}^{(n)}})$, which makes it possible to automatically combine the various constraints associated with the landmarks and to use the dynamical controller as previously determined.

5.3.3   Handling of inverse kinematics

Velocity is first estimated by numerical integration as in (5.3). To find a controller for the robot that reproduces a desired path in Cartesian space, we here utilize an approach based on a damped least square inverse kinematics solution that locally optimizes a cost function in the null space of the Jacobian matrix (see Sect. 4.1). At each iteration, we compute

$$\dot{\theta} = J^{\dagger}(\theta) \, \dot{x} + N(\theta) \, (\hat{\theta} - \theta),$$

where $J^{\dagger}(\theta)$ is the pseudoinverse of the Jacobian matrix and $N(\theta)$ is the null space projection matrix according to

$$N(\theta) = I - J^{\dagger}(\theta)J(\theta).$$

Here, $\hat{\theta}$ represents the desired posture that the robot should follow as best as possible provided that the kinematic chain allows some redundancy, which is the case with the robot's arm in the considered example (3D positions achieved with 4 DOFs). This desired posture is computed by generalizing over the joint angle trajectories collected during the demonstrations. This is done by encoding the joint probability $\mathcal{P}(\theta, x)$ in a GMM and by at each iteration retrieving generalized joint angles $\hat{\theta}$ through an estimation of $\mathcal{P}(\theta|x)$ with GMR, thus rendering it possible to estimate the most likely postures $\hat{\theta}$ along the motion. Such an approach allows the design of simple controllers for tasks that mainly focus on following paths in Cartesian space with the aim of manipulating objects with a redundant kinematic chain. In essence, the method follows the same approach as the generic framework described in previous chapters, but gives priority to constraints in task space by considering constraints in joint space at only a secondary level.

5.4   EXPERIMENTAL RESULTS

For the first experiment, encoding and reproduction results in Cartesian space are presented in Figures 5.11 and 5.12. We can see that the proposed method correctly deals with perturbations and with a variety of initial configurations. The influence of the VITE parameters when computing the controller allows a tuning to the desired smoothness in order to return to a learned situation when reproducing the skill. Figure 5.13 presents the results in joint space,

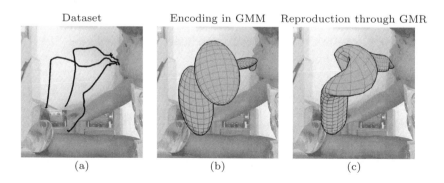

Dataset   Encoding in GMM   Reproduction through GMR

(a)   (b)   (c)

Figure 5.11 Encoding results for the first experiment. (a) Three demonstrations of the skill in the frame of reference of Robota's mouth by moving the robot and the initial position of the spoon from one demonstration to the other. (b) Gaussian Mixture Model of 4 Gaussian distributions encoding the skill in *Robota's* frame of reference. (c) A continuous model retrieved through Gaussian Mixture Regression.

From different initial positions   By moving landmark $o^{(1)}$ By modulating dynamical system

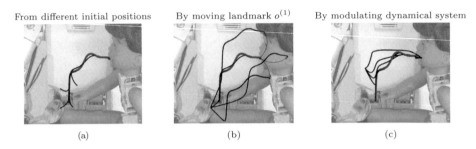

         (a)   (b)   (c)

Figure 5.12 The reproduction results for the first experiment where the generalized trajectory is represented with a thick line and where the reproduced trajectories are represented with a thin line. (a) The reproduction by starting from different initial positions. (b) Various reproduction attempts by starting from the same initial point and by moving Robota's head in varying directions while HOAP-3 moves its right arm. (c) A modulation of the reproduction behavior by varying the parameters of the dynamical systems.

where the inverse kinematics solution also takes into account the demonstrated trajectories in joint angle space.

For the second experiment, Figure 5.14 presents the encoding results and Figure 5.15 presents the reproduction data. One can see that the robot automatically combines the two sets of constraints (associated with its own mouth as well as with Robota's mouth) to find an optimal trade-off probabilistically satisfying the variabilities observed during the demonstrations.

The time used to train the model is not part of the robot control loop and is thus not a limitative factor. The training of the GMM takes between one and five seconds for the small number of demonstrations considered. When taking into account reproduction, with the weight computed as in (5.4), the retrieval process takes approx. 2 milliseconds for the simple arm considered here (with a non-optimized implementation in C++ on a standard PC running *Linux*). If the robot kinematics is more complex or if the controller requires a very high control rate, a practical and computationally efficient solution consists in approximating the weight in (5.4) as

$$h_i(\xi_t) = \frac{\mathcal{N}(\xi;\ \mu_i,\ \Sigma_i)}{\sum_{j=s}^{s+1} \mathcal{N}(\xi;\ \mu_j,\ \Sigma_j)} \quad \left|\ \begin{matrix} \forall i \in \{s, s+1\}, \\ s = \operatorname*{arg\,max}_{i \in \{1,...,K\}} \alpha_{i,t}, \end{matrix} \right. \qquad (5.5)$$

where s represents the most probable current state of the HMM that explains the current partial observation.

A direct extension of the system proposed in this chapter consists in encoding the joint distribution $\mathcal{P}(x, \dot{x}, \ddot{x})$ in GMM and retrieving at each time step an acceleration command by estimating $\mathcal{P}(\ddot{x}|x, \dot{x})$ through GMR (here, $\xi^{\mathcal{I}} = [x, \dot{x}]$ and $\xi^{\mathcal{O}} = \ddot{x}$). As described in Hersch, Guenter, Calinon, and Billard (2008), this approach was utilized to combine an acceleration-based model of the skill with a mass-spring-damper system, used as an attractor, to the end of the motion (e.g., when reaching for an object). By modulating this dynamical

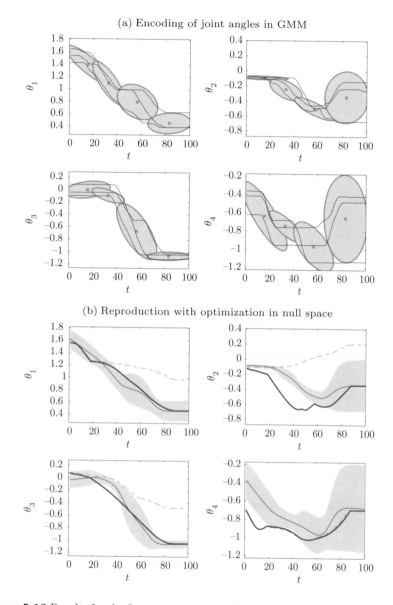

Figure 5.13 Results for the first experiment in joint angle space. (a) Encoding of the various demonstrations in GMM. (b) The generalized trajectory is represented with a thin solid line and its associated surrounding envelope. The reproduced trajectory is represented with a thick line. Primarily, this trajectory satisfies the constraints in task space. The constraints in joint space are then satisfied through projection in the null space of the Jacobian matrix. For the sake of comparison, the solution found by least-norm inverse kinematics is represented with a dashed line. In the second graph, a joint angle limit at $\theta_2 = 0$ appears. We see that, by optimizing the controller to follow the generalized joint angle trajectory, the robot correctly avoids this joint limit.

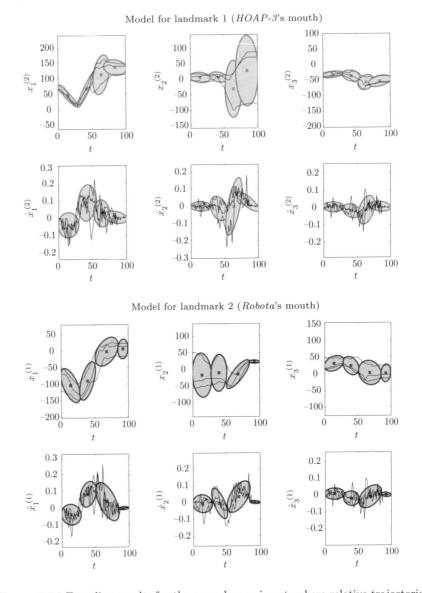

Figure 5.14 Encoding results for the second experiment, where relative trajectories with respect to the two landmarks are encoded in two GMMs of 4 Gaussians. The two rows show position and velocity variables along the task, where the forms of the Gaussian distributions display that some parts of the motion are more constrained than others. The covariance matrices for landmark 1 are smaller in the first part of the motion, signifying that the observed demonstrations were consistent for this part of the skill and that they should probably be reproduced in this specific way. To reproduce the skill, it is important for the robot to take into account the constraints associated to its mouth at the beginning of the skill. For landmark 2, the task is more constrained at the end of the motion, which is depicted by smaller covariance matrices at the end of the task (particularly for variable x_2).

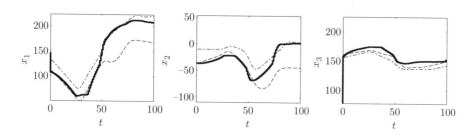

Figure 5.15 Reproduction results for the second experiment. The final reproduction is represented with a thick line, and the two sets of constraints, i.e., the dashed line for HOAP-3's mouth and the dash-dotted line for Robota's mouth, have been superimposed to show that the robot tends to satisfy the constraint related to landmark 1 first (i.e., reaching its own mouth) before smoothly switching to the second constraint (i.e., reaching the other robot's mouth).

system with the acceleration-based model learned during the demonstrations, the robot first followed the observed behavior and then reached its target by switching smoothly from one controller to the other.

The main advantage of an acceleration-based controller $\mathcal{P}(\ddot{x}|x, \dot{x})$ over one based on velocity $\mathcal{P}(\dot{x}|x)$ is that it allows the learning of a motion presenting more complex dynamics. On the other hand, as the acceleration signal is more complex than the velocity and position signals, the joint probability distribution may be represented too coarsely by the Gaussian distributions if the skill presents subtleties that are difficult to model. For a complex motion, a compact encoding of the skill based on acceleration may indeed not satisfy the precision required to follow a desired path. If the motion to learn involves continuous changes of the attractor's position and dynamics that are mainly described by a specific path to follow, the use of the simpler velocity model $\mathcal{P}(\dot{x}|x)$ thus remains an appealing and computationally efficient solution.

CHAPTER 6

TRANSFERRING SKILLS THROUGH ACTIVE TEACHING METHODS

So far, we have considered experiments where the demonstration phase was separated from the reproduction phase. This chapter discusses the importance of adding social components to the teaching paradigm, not only from the perspective of designing user-friendly interfaces, but also from the point of view of designing efficient teaching scenarios. Indeed, we demonstrate here that the transfer of skills can take advantage of several psychological factors that the user might encounter during teaching.

The teaching paradigm is complex and involves a combination of social mechanisms. In order to transfer new skills to humanoid robots, several directions in Human-Robot Interaction (HRI) were adopted to benefit from the shape, human likeness and interaction capabilities of these robots. Kahn, Ishiguro, Friedman, and Kanda (2006) and MacDorman and Cowley (2006) explored which components of the interaction that were required to efficiently learn new skill through the teaching interaction. Salter, Dautenhahn, and Boekhorst (2006) reviewed possible interactions through speech, gestures, imitation, observational learning, and instructional scaffolding. To these were added physical interaction such as molding or kinesthetic teaching. B. Lee, Hope, and Witts (2006) explored the use of behavioral synchronization as a powerful communicative experience between a human user and a robot, and demonstrated through experiments that in human dyadic interactions, close relationship partners are more engaged in joint attention to objects in the environments and to their respective body parts than strangers. Thomaz et al. (2005) presented HRI experiments in which the user could guide the robot's understanding of the objects in its environment through facial expressions, vocal analysis, detection of shared attention and by using an affective memory system. Rohlfing, Fritsch, Wrede, and Jungmann (2006) highlighted the importance of multimodal cues to reduce the complexity of human-robot skill transfer and considered multimodal information as an essential element to structure the demonstrated tasks. Riley, Ude, Atkeson, and Cheng (2006) highlighted the importance of an active participation of the teacher not only to demonstrate a model of expert behavior but also to refine the acquired motion by vocal feedback. Yoshikawa, Shinozawa, Ishiguro, Hagita, and Miyamoto (2006) suggested that the teacher should no longer be considered solely as a model of expert behavior but should become an active participant in the learning process.

The present chapter adopts the perspective that the transfer of skills can take advantage of several psychological factors encountered by the user during the teaching interaction. Similarly to Yoshikawa et al. (2006), we consider the role of the teacher as one of the most important key components for an efficient transfer of the skill. The teaching interaction is then designed to allow the user to become an active participant in the learning process (and not only a model of expert behavior). By taking inspiration from the human tutelage paradigm, Breazeal et al. (2004) showed that a socially guided approach could improve both the human-robot interaction and the machine learning process by taking into account human benevolence. In their work, they highlighted the role of the teacher in organizing the skill into manageable steps and maintaining an accurate mental model of the learner's understanding. However, the tasks considered were mainly built on discrete events organized hierarchically. Our work shares similarities with theirs in terms of the tutelage paradigm, but we focus on learning continuous motion trajectories and actions on objects at a trajectory level instead of considering discrete events. Saunders et al. (2006) provided experiments where a wheeled robot is teleoperated through a screen interface to simulate a *molding* process, i.e., by letting the robot experience sensory information when exploring its environment through the teacher's support. Their model used a memory-based approach in which the user provides labels for the different components of the task to hierarchically teach high-level behaviors. Our work adopts similar ideas but follows a model-based approach, where teleoperation is replaced by the more user-friendly means of directly moving the robot's arms while the robot records sensory information through its motors encoders.

6.1   EXPERIMENTAL SETUP

These experiments consider teaching interactions for which various modalities were used to produce the demonstrations (Fig. 6.1). Three experiments were performed based on the incremental learning approach with the direct update method presented in Section 2.6.3, allowing the teacher to gradually visualize the results of his/her demonstrations. The use of motion sensors (see Sect. 3.1) allows the user to demonstrate high-dimensional movements, while kinesthetic teaching (see Sect. 3.2) provides a means of supporting the robot in its reproduction of the task (Fig. 6.1). By using scaffolds, the user provides support to the robot by manually articulating a decreasing subset of motors. The scaffolds progressively fade away while the user finally lets the robot perform the task on its own, allowing the robot to experience the skill independently. Therefore, the teacher uses the scaffolding process to permit the robot to gradually

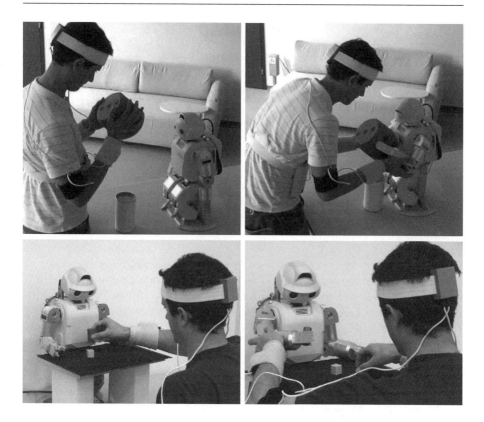

Figure 6.1 Various modalities are used to convey the demonstrations and scaffolds required by the robot to learn a skill. The user first demonstrates the whole movement using motion sensors (*left*) and then helps the robot refine its skill through kinesthetic teaching (*right*), i.e., by *embodying* the robot and accompanying it through the motion. *Top:* Learning bimanual gestures. *Bottom:* Learning to stack objects.

generalize the skill in an increasing range of contexts.[1] Knowing this, it may be important to reproduce the acquired skill after each demonstration to help the teacher prepare the following one according to the outcome. This shares similarities with the human process of refining the understanding of a task through practice and corrective feedback, by incrementally combining new information with previous understanding. It thus suggests the use of incremental learning approaches when teaching humanoid robots, allowing them to extend, refine and elaborate their understanding of the skill.

The robot builds a model of the task constraints by observing the user as he demonstrates a manipulation skill using different initial positions of the objects. After each demonstration, the robot reproduces a generalized version

---

[1] *Scaffolding* is here used as a metaphor for construction sites where scaffolds are used to build an infrastructure and where the scaffolds are progressively withdrawn as the infrastructure gets more and more consolidated, up to the point where the construction is finished.

of the task by probabilistically combining the various constraints encapsulated in the GMMs. Watching the robot reproduce the task after each demonstration helps the teacher evaluate the reproduction and generalization capabilities of the robot. The teacher can thus detect the portions of the task that require refinement or that have not been correctly *understood* by the robot, helping the teacher prepare further demonstrations. The model is incrementally refined after each demonstration, and the user decides to stop the interaction when the robot has correctly learned the skill.

By extracting the variations and correlations through GMR (Sect. 2.4), one can, at each time step along the trajectory, detect what the relevant variables composing the task are and how they are correlated. For some tasks, relevant and irrelevant variables can be clearly separated (Alissandrakis, Nehaniv, Dautenhahn, & Saunders, 2005), defined a priori, or be constant throughout the task. Here, the most general case is considered, in which a variety of constraint levels are allowed, and where the constraints can change during the skill. Our belief is that representing the task constraints in a binary manner (relevant versus irrelevant features) is inappropriate for continuous movements. Indeed, some goals require specific precisions, i.e., they can be described with varying degrees of invariance. For example, the movement used to drop a lump of sugar in a cup of coffee is more constrained than the movement to drop a bouillon cube in a large pan.

Trajectory constraints are defined for each object considered in the scenario, thereby allowing the system to simultaneously model gestures and actions on objects, using the direct computation method described in Section 2.5.1. This probabilistic representation of the trajectory constraints on each object is then used to reproduce the task under new conditions, i.e., with in positions objects that have not been used to demonstrate the skill. Extracting the constraints of a movement is important for determining which parts are relevant, which ones allow variability, and what kind of correlations there are among the variables. Similarly to humans, an efficient human-robot teaching process should thus encourage variability in the demonstrations provided to the robot, e.g., varied practiced conditions, varied demonstrators or varied exposures. Thus, the extraction of variability and correlation information may allow the robot to use its experience in changing environmental conditions, i.e., where the task demands may become modified during the skill. Using GMMs, this can be set-up in an adaptive way without drastically increasing the complexity of the system when new experiences are provided.

The experiment is conducted using the *HOAP-3* robot of which only the 16 DOFs of the upper torso are required in the experiments. Two webcams in its head are used to track objects in 3D Cartesian space based on color information as presented in Section 4.2. The objects to track are predefined in a calibration phase. Alternatively, the initial positions of the objects can also be recorded by a *molding* process where the teacher grabs the robot's arm, moves it toward the object and positions the robot's palm around the object. When the object touches its palm, the robot *feels* the object by way of a force sensor. It then briefly grasps and releases the object while registering its position in 3D Cartesian space.

Two separate modalities are used to convey the demonstrations (Fig. 6.1). First, motion sensors attached to different body parts of the user are employed (see Sect. 3.1). The user's movements are recorded by 8 *X-Sens* motion sensors attached to the torso, upper-arms, lower-arms, hands (at the level of the fingers) and back of the head. The motor encoders of the robot are then used to record information while the teacher moves the robot's arms. The teacher selects which motors to manually control by slightly moving the corresponding motors before the reproduction starts. The selected motors are set to passive mode which permits the user to freely move the corresponding degrees of freedom while the robot executes the task.

Three experiments are presented to show that the method is generic and can be used with datasets representing different modalities. For the first experiment, the robot collects the joint angle trajectories of the two arms of the user, which are the lowest-level data that the robot can collect. We show through this experiment that the system can efficiently encode high-dimensional data by projecting them in a latent space of motion. For the second experiment, the robot collects the 3D Cartesian path of the right hand (using direct kinematics). The aim is to show that manipulation skills can be learned by the system and generalized to various initial positions of objects. The third experiment extends this manipulation skill to the use of a variety of objects, whose specific properties are learned through demonstrations. The first two experiments were presented in Calinon and Billard (2007d) and the last experiment was described in Calinon and Billard (2007a).

6.2   EXPERIMENTAL RESULTS

6.2.1   *Experiment 1:* learning bimanual gestures

This experiment shows how a bimanual skill can be incrementally taught to the robot in joint angle space using observation and scaffolding. The task consists in grasping and moving a large foam cube (Fig. 6.1 and 6.2). Starting from a resting pose, the left arm is first moved to touch the left side of the cube, with the head following the motion of the hand. This is followed by a symmetrical gesture being performed with the right arm. When both hands grasp the object, it is lifted and pulled back on its base, with the head turned toward the object (Fig. 6.2). The user, wearing the motion sensors, performs the first demonstration of the complete task, which allows him/her to demonstrate the full gesture by simultaneously controlling 16 joint angles. These joint angles are then projected onto a subspace of lower dimensionality through PCA. After observation, the robot reproduces a first generalized version of the motion. This motion is then refined by kinesthetically helping the robot perform the gesture, or in other words, by physically moving its limbs during the reproduction attempt. The gesture is partially refined by guiding the desired DOFs while the robot controls the remaining DOFs. Based on this method, the teacher can only move a limited subset of DOFs by using his/her two arms, which signifies that the user can move the two shoulders and the

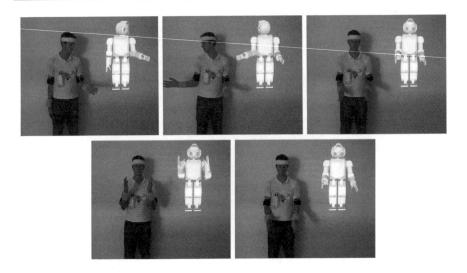

Figure 6.2 Illustrations of the use of motion sensors to acquire gestures from a human model. A simulation of the robot is projected behind the user to show how the robot collects the joint angles during the process. The gestures are similar to those from *Experiment 1*, except that in these snapshots, the user does not face the robot and only mimics the grasping of the object.

two elbows of the robot simultaneously, but that the remaining DOFs (head, wrists and hands) are controlled by the robot. The robot thus reproduces the movement while recording the movement of the limbs controlled by the user.

Results of the experiment are presented in Figures 6.3 and 6.4, where the resulting paths of the hands resemble those demonstrated by the user (see Fig. 6.2), even if training is performed in a subspace of motion where joint angle trajectories have been projected. The system finds five principal components and five Gaussians to efficiently represent the trajectories in latent space. After the first demonstration of the movement while wearing motion sensors, the robot can only reproduce a smoothed version of the joint angles produced by the user. Due to the difference in size between the user's and robot's bodies, whereas the cube does not change, the robot hits the cube when trying to reproduce the skill. This is the result of the robot moving its left hand first and tipping the cube over before moving its right hand. Observing this, the teacher progressively refines the model by providing appropriate scaffolds, i.e., by controlling the shoulders and the elbows of the robot while reproducing the skill so that it may grasp the cube correctly. In the third reproduction attempt, the robot lifts the cube awkwardly. In the sixth attempt, the robot skillfully reproduces the task by itself (see Fig. 6.4). The user thus decides to stop the teaching process at this point.

6.2.2   *Experiment 2:* **learning to stack objects**

This experiment shows how the system can learn manipulation skills incrementally in 3D Cartesian space. Thanks to the teacher's support, the robot extracts

First attempt   Third attempt   Sixth attempt

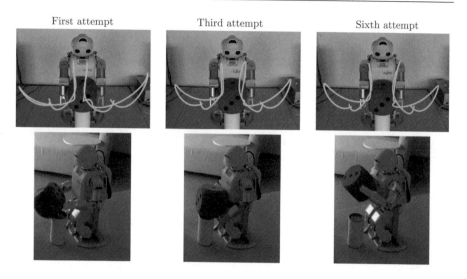

Figure 6.3 The reproduction of the task after the first, third and sixth demonstration. In the first attempt, the robot hits the cube when approaching it. In the third attempt, the robot's skill gets better but the grasp is still unstable. In the sixth attempt, the cube is correctly grasped. The trajectories of the hands are plotted in the first row, and the corresponding snapshots of the reproduction attempts are represented in the second row.

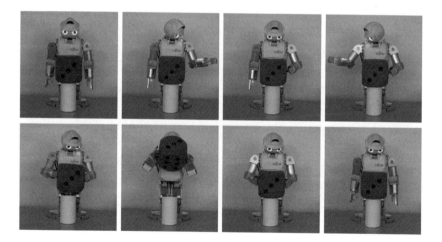

Figure 6.4 Snapshots of the sixth reproduction attempt.

and combines constraints related to different objects in the environment. The aim of the task is to grasp a red cylinder with the right hand and place it on a yellow cube (see Fig. 6.5). This experiment also shows that teaching manipulation skills can be achieved through a scaffolding process where the user gradually highlights the *affordances* of various objects in its environment and

Figure 6.5 The reproduction of actions on objects. The purpose of the task is to grasp the red cylinder and place it on top of the yellow cube. The first snapshot shows the 3D Cartesian frame of reference used in the experiment.

the *effectivities* of the body required to perform grasping and moving actions on these objects, drawing insights from experiments in developmental psychology where a caregiver teaches manipulation skills to a child (Zukow-Goldring, 2004). In the proposed setup, the robot simultaneously learns the *affordances* of the two objects (i.e., the properties of the red cylinder permit it to be stacked on the yellow cube) and the associated *effectivities* (i.e., how the robot should use its body to grasp and place the cylinder on top of the cube without hitting any object).

The specific constraints related to the two objects are extracted by varying the demonstrations provided to the robot, i.e., by starting with different initial positions of the objects. Thus, the efficiency of the reproduction relies mainly on the ability of the teacher to provide appropriate scaffolds when demonstrating the task, or in other words by using sufficient and appropriate variability across the demonstrations. In doing so, the system is able to generalize and reproduce the skill in situations that have not been used to demonstrate the task. After each demonstration, the robot tries to reproduce the skill in two different situations. This helps the teacher evaluate how well the robot can generalize under altered conditions.

The system simultaneously learns the 3D Cartesian trajectories relative to the various objects in the environment and the actions used to move these objects, as suggested in Section 2.5.1. In Section 4.1, several Jacobian based inverse kinematics methods for dealing with the robot's arm redundancies were presented. Here, a much simpler strategy, similar to that of Shimizu, Yoon, and Kitagaki (2007), is used. A posture is defined by the position of the right hand in a 3D Cartesian space and by an additional angular parameter α determining the elevation of the elbow (angle formed by the elbow and a vertical plane, see Fig. 6.7). For each demonstration, the hand path and a trajectory of angle α fully describe the gesture. Generalized versions of the hand paths and of the α trajectories are then used to compute by geometrical inverse kinematics the joint angle trajectories required to control the robot.

In this experiment, $n = 6$ demonstrations of the task are performed where each trajectory is rescaled to $T = 100$ time steps (after six demonstrations, the user considered that the robot had understood the task correctly). The total number of observations could thus be given by $N = nT$. For each demonstration $i \in \{1, \ldots, n\}$, the collected data consists of the initial positions of the M objects $\{o_i^{(h)}\}_{h=1}^M$ and of the set of variables $\{x_{i,j}, \alpha_{i,j}\}_{j=1}^T$ corresponding to the absolute right hand paths and to the evolution of the parametric angle α. After each demonstration, the trajectories of the right hand relative to the M objects are computed with

$$x_{i,j}^{(h)} = x_{i,j} - o_i^{(h)} \quad \left| \begin{array}{l} \forall h \in \{1, \ldots, M\} \\ \forall i \in \{1, \ldots, n\} \\ \forall j \in \{1, \ldots, T\}. \end{array} \right.$$

The constraints of the task are extracted from $\{x^{(1)}, \ldots, x^{(M)}, \alpha\}$ to which encoding and generalization are applied separately. By using two objects, a generalized version of the trajectories is thus given by $\{\hat{x}^{(1)}, \hat{x}^{(2)}, \hat{\alpha}\}$ and the associated covariance matrices $\{\hat{\Sigma}^{(1)}, \hat{\Sigma}^{(2)}, \hat{\Sigma}^\alpha\}$.

Results for the encoding, generalization and extraction of constraints are presented in Figure 6.6 (see also Fig. 6.5 for the x_1, x_2, x_3 directions). For the first object, the trajectory $\hat{x}^{(1)}$ can be seen to be highly constrained when grasping the object between time steps 20 and 40 (a tight envelope representing the constraints around the generalized trajectory). The constraints were tighter for the first two variables $\hat{x}_1^{(1)}$ and $\hat{x}_2^{(1)}$, defining the movement with respect to the surface of the table. This was consistent with the shape of the object to grasp (a cylinder placed vertically on the table). Indeed, the form and orientation of the object enabled it to be grasped with more variability on the third axis x_3, defining the vertical movement. When observing the constraints associated with $\hat{x}_3^{(1)}$, one observes that between time steps 40 and 70, the generalized trajectories of the hand holding the object followed bell-shaped paths, which was consistent with the provided demonstrations. For the second object, the trajectory $\hat{x}^{(2)}$ was highly constrained at the end of the task, when placing the cylinder on top of the cube.

The reproduction of the task in two new situations is presented in Figure 6.9. After the first demonstration, the system was unable to generalize the skill (a similar motion was used for the two reproduction conditions). After the third demonstration, the constraints were already roughly extracted. The robot grasped the cylinder correctly but still had some difficulty in placing it on the cube. By watching the robot reproduce the skill in the two situations, the teacher detected, during the third reproduction attempt, that the second part of the task was not fully understood. The teacher then provides appropriate scaffolds by kinesthetically demonstrating the task with an increased variability in the initial positions of the cube. Thus, the constraint of placing the object on top of the cube becomes more salient, and at the sixth attempt, the various constraints are finally fulfilled for the different reproduction conditions; namely, grabbing the cylinder, moving it with an appropriate bell-shaped movement, and placing it on top of the cube.

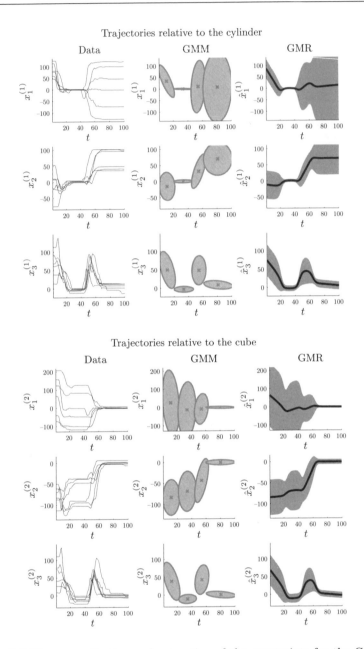

Figure 6.6 The generalization and extraction of the constraints for the Cartesian trajectories relative to the two objects, after six demonstrations of the skill. *Left:* The six demonstrations observed consecutively for each object. *Middle:* The *Gaussian Mixture Model* (GMM) of four components used to incrementally build the model; *Right:* A representation of the generalized trajectories and associated constraints extracted by *Gaussian Mixture Regression* (GMR).

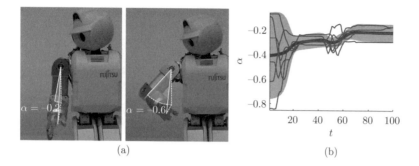

(a)   (b)

Figure 6.7 (a) The angle parameter α used to define the gesture through geometrical inverse kinematics. (b) The extraction of the generalized trajectory (*thick line*) and associated constraints (*shaded envelope*) after six observations (*thin line*) of angle parameter α.

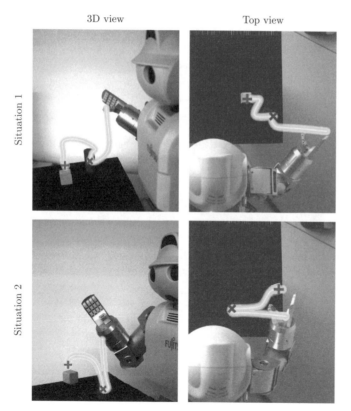

Figure 6.8 The reproduction of the task for 2 situations and after 6 demonstrations of the skill. Despite that the initial positions of the two objects have not been observed during the demonstrations, the robot is able to generalize and fulfill the task requirements, namely to grasp the cylinder and place it on top of the cube.

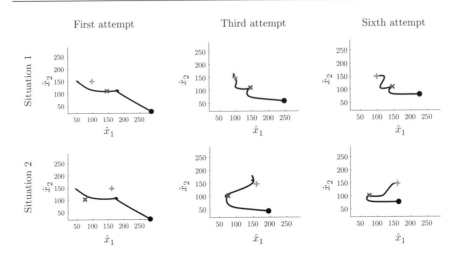

Figure 6.9 The reproduction of the task after the first, third, and sixth demonstrations for two new initial positions of the objects. The solid line represents the hand path, with the starting position marked with a dot. The cross and a plus sign represent the cylinder and the cube, respectively.

6.2.3 *Experiment 3:* learning to move chess pieces

This experiment uses a chess game to demonstrate how a robot can learn manipulation skills in 3D Cartesian space. The chess paradigm was also used by Alissandrakis, Nehaniv, and Dautenhahn (2002) to explore the correspondence problem in imitation learning, i.e., to know how to reproduce a particular motion on the chessboard by considering different chess pieces (different *embodiments*). With the teacher's support, the robot extracts and combines constraints related to various objects in the environment. The setup of the experiment is presented in Figure 6.10. Six consecutive kinesthetic demonstrations are provided demonstrating how to grasp a Rook, a Bishop or a Knight from a chessboard and how to bring the grasped chess piece to the King of the adversary, see Figures 6.11, 6.12 and 6.13. Each demonstration is rescaled to $T = 100$ datapoints. Three models are created for the Rook, the Bishop and the Knight, and the setup is simplified by using only two chess pieces at the same time (attentional scaffolding can be used to permit the robot to recognize only these two chess pieces). The Rook is moved exclusively in a forward linear direction, the Bishop is moved solely in a forward-left diagonal direction, and the Knight is moved only two squares forward followed by one square to the left, forming an inverse 'L' shape.[2]

Similarly to the previous experiment, the robot simultaneously learns the *affordances* of the objects (i.e., the Rook/Bishop/Knight are moved in separate

---

[2] Note that the basic rules remain the same and that the directions only depend on the considered frame of reference.

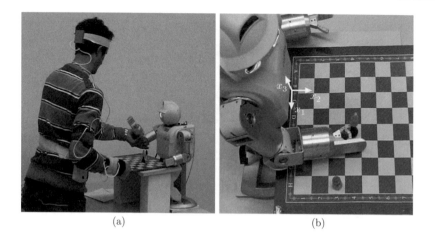

(a)   (b)

Figure 6.10 The experimental setup. (a) A kinesthetic demonstration. (b) The reproduction and frame of reference.

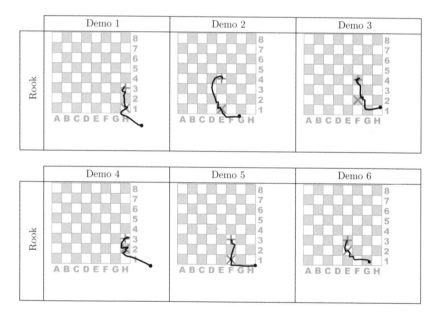

Figure 6.11 Trajectories for the 6 consecutive kinesthetic demonstrations for the Rook (only x_1 and x_2 are represented, corresponding to the plane of the chessboard). The cross and the plus sign represent respectively the Rook and the King.

ways to be brought to the King) and the associated *effectivities* (i.e., how the robot should use its body to grasp and bring the Rook/Bishop/Knight to the King without hitting another object). After each demonstration, the robot tries to reproduce the skill with the chess pieces placed in a new situation that has not been observed during the demonstrations. Thus, the user can test the ability of the robot to generalize the skill over a variety of situations.

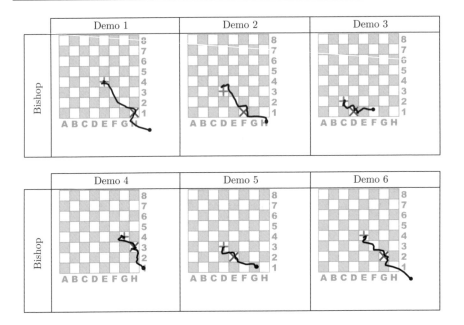

Figure 6.12 Trajectories for the 6 consecutive kinesthetic demonstrations for the Bishop. The cross and the plus sign represent respectively the Bishop and the King.

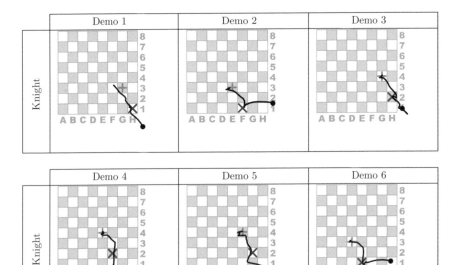

Figure 6.13 Trajectories for the 6 consecutive kinesthetic demonstrations for the Knight. The cross and the plus sign represent respectively the Knight and the King.

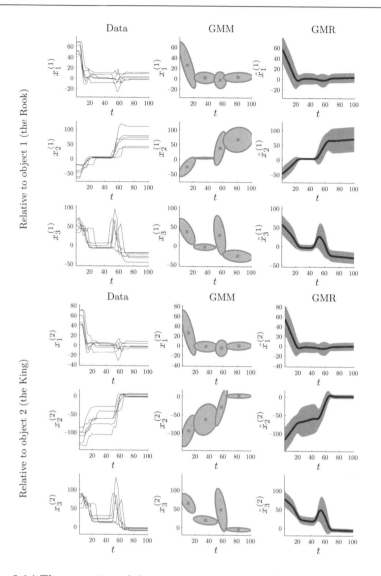

Figure 6.14 The extraction of the constraints relative to the Rook (1st object) and to the King (2nd object). In each cell, the first column represents the right hand paths collected in a 3D Cartesian space, the second column represents the GMM encoding of the data and the last column represents the GMR process used to extract the generalized trajectories with associated constraints. The parts that show a thin envelope around the generalized trajectory are locally highly constrained, while the parts presenting large envelopes allow a loose reproduction of the trajectory.

The constraints extracted after the sixth demonstration are presented in Figures 6.14, 6.15 and 6.16, displaying the displacement constraints with respect to the first object (either the Rook, the Bishop or the Knight) and to the second object (the King). For the Rook, Figure 6.14 shows that the

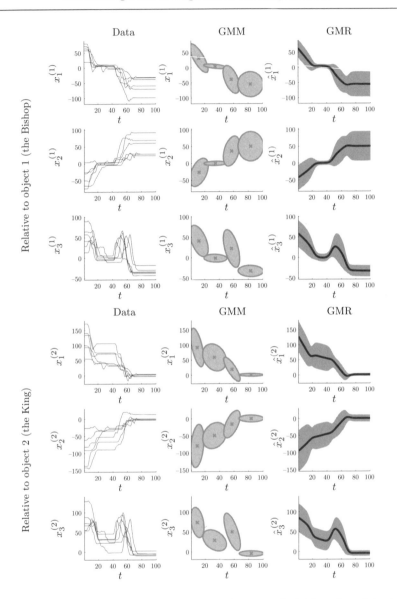

Figure 6.15 The extraction of the constraints relative to the Bishop (1st object) and to the King (2nd object).

trajectory is highly constrained for $x_1^{(1)}$ from time step 30 while the hand grasps the Rook. The Rook is then moved in a straight line, i.e., the direction in x_1 remains constant after grasping of the Rook. However, its final position can change in amplitude, i.e., the Rook is moved along a straight line but its final position on this line can vary, which is reflected by the constraints extracted for $x_2^{(1)}$ (larger envelope after time step 70). For the Bishop, Figure 6.15 shows that the generalized trajectories relative to Object 1 follow a diagonal. The

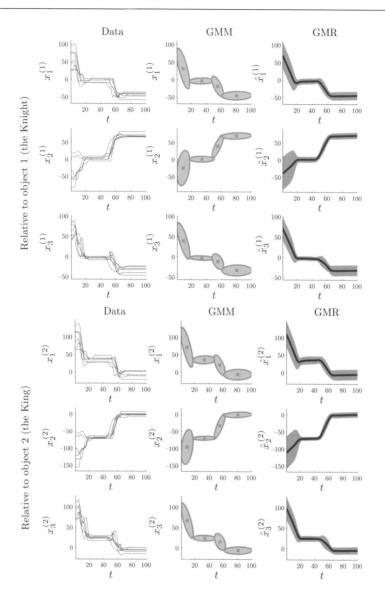

Figure 6.16 The extraction of the constraints relative to the Knight (1st object) and to the King (2nd object).

direction is highly constrained but the final position is not. For the Knight, we can see from Figure 6.16 that the generalized trajectories relative to Object 1 are more constrained. Indeed, for a given initial position of the Knight, only one final position is allowed in the proposed setup. This is reflected by the constraints for $x_1^{(1)}$ and $x_2^{(1)}$ (and complementary for $x_1^{(2)}$ and $x_2^{(2)}$), where the path with respect to the initial position of the Knight is highly constrained (the followed path is quasi-invariant across all demonstrations). For the three chess pieces, the constraints for the vertical axis $x_3^{(1)}$ share similarities, demonstrating that the user grasps the chess piece from above, and displaces it according to a bell-shaped trajectory in a vertical plane. For each chess piece, the constraints for the first object are correlated with the constraints for the second object. Indeed, the positions of the two objects have important dependencies for the skill (i.e., it is important to reach the King of the opponent with the chess piece).

By considering the constraints relative to the two objects, the reproduction of the hand path is then computed for new initial positions of the objects. To do so, the absolute constraint for each object is computed by adding the new initial position of each object to the corresponding relative constraint (represented as centers and associated covariance matrices along the motion). The hand path used for reproduction is then computed by multiplying, at each time step, the two Gaussian distributions characterizing the absolute constraints for the two objects. This hand path is converted to joint angles through a geometrical inverse kinematics algorithm, parameterized by the generalized version of the angle α between the elbow and a vertical plane, which allows to reproduce naturally-looking gestures by providing an additional constraint to the redundant inverse kinematics problem. Figure 6.17 shows the generalized trajectories and associated constraints on the angle parameter α (see also Fig. 6.7). For the three chess pieces, the gestures used to reach for the chess piece share similarities. Indeed, the α trajectories start with a negative value and progressively converge to zero. This means that the elbow is first elevated outward from the body and is then progressively lowered, approaching the body until the arm and forearm are almost in a vertical plane. This allows the user (and the robot) to approach the chess piece carefully without hitting other pieces. Indeed, when experiencing the skill together with the robot through a scaffolding process, the user quickly notices that when the robot is close to the chess piece, its elbow has to be lowered to grasp the chess piece correctly. This is mainly due to the missing backdrivable motor at the level of the wrist which constrains the grasping posture of *HOAP-3* parameterized by an angular value α close to zero.

When the chess piece is grasped, we see that either of two movement strategies are adopted depending on the path to follow. As the Rook is moved

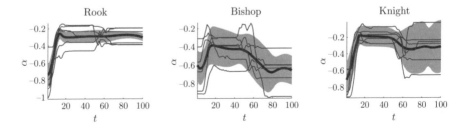

Figure 6.17 The extraction of the constraints on the gesture (modeled by the angle parameter α) used to move the 3 chess pieces.

forward, the angular configuration does not need to be changed after grasping. As the Bishop is moved diagonally (forward-left), the user helps the robot adopt a correct posture in order to avoid hitting its own body when performing the draw (i.e., learning *effectivities*). This is reflected by the decreasing negative value of angle α. A similar strategy is employed for the Knight, but the amplitude of change in angle α is lower due to the shorter path followed by the Knight. We also see that there is a slight tendency to first keep the arm in a vertical plane (α close to zero), and to finally use a posture with a slight elevation of the elbow (negative α value). This behavior is probably due to the inadvertent decomposition of the 'L' shape demonstrated by the user when kinesthetically helping the robot displace the Knight.

Figure 6.18 presents several reproduction attempts for initial configurations that have not been observed by the robot during the teaching process. We see that the ability of the robot increases, i.e., each demonstration helps the robot refine its model of the skill as well as its ability to generalize across different situations.

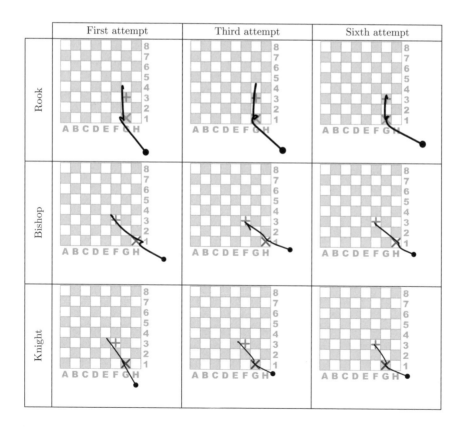

Figure 6.18 The reproduction of the task for the Rook, Bishop and Knight after observation of the first, third and sixth demonstration. The cross and the plus sign represent respectively the chess piece to grasp and where the grasped piece should be moved.

Through these experiments, the importance of designing a PbD framework that takes into account the complete interaction process instead of merely the learning system is highlighted. The active role of the user in the teaching process is emphasized by putting him/her in the loop of the robot's learning. Three experiments were presented to illustrate the use of an incremental and interactive PbD framework to teach manipulation skills to a humanoid robot.

6.3   ROLES OF AN ACTIVE TEACHING SCENARIO

We discuss here the advantages of considering teaching scenarios that put the user in the loop of the robot's learning. Several social learning insights from a multitude of research fields are presented, and have been used as hints to design the experiments portrayed in this chapter in order to give the user as much help as possible in providing relevant data for the robot. We discuss how PbD benefits from active and incremental teaching methods, and how the pedagogical skills of the user can contribute to an efficient transfer of the skill.

To design efficient teaching systems and appropriate benchmarks, it is important to take psychological and social factors into consideration. Even if, as presented in Section 2.2.1, the learning efficiency can be quantitatively measured , i.e., by determining how well the system reproduces and generalizes the skill with respect to a specific dataset, the performance also depends directly on the quality of the dataset and on the teaching abilities of the user. It is thus important to consider the skill transfer process, not only at the algorithmic level but also at the user level, by designing human-robot interactive scenarios that make efficient use of the teaching abilities of the user. Several insights covering psychology, pedagogy, developmental sciences, sociology and sports science are now presented.

6.3.1   Insights from psychology

Considering the robot as a peer learner and watching the evolution of its understanding of the skill are important psychological factors for a successful interaction. By drawing parallels with a caregiver-child interaction (and the associated human social aptitude at transmitting culture), we see that, by watching the evolution and the outcomes of teaching, the teacher can feel psychologically more implicated in the teaching teamwork. In psychology, benchmarks such as autonomy, moral accountability and reciprocity have been proposed to evaluate Human-Robot Interaction (HRI) setups (Kahn et al., 2006). For teaching applications, the robot's capacity to generalize over several contexts depends on the number of demonstrations it has been provided with, but more importantly on the *pedagogical* quality of these demonstrations (gradual variability of the situations and exaggerations of the key features to reproduce). To succeed, it is therefore crucial that the teacher feels implicated in the teamwork.

Indeed, an essential feature of human beings is our natural ability and desire to transmit skills and knowledge to the next generation. This transfer problem is complex and involves a combination of social mechanisms such as speech, gestures, imitation, observational learning, instructional scaffolding, as well as physical interaction such as molding or kinesthetic teaching (Salter et al., 2006). The humanoid shape of the robot and the humanlike properties of the teaching system aim at helping the teacher consider the robot as a peer. As noted by MacDorman and Cowley (2006), a humanoid robot is the best equipped platform for reciprocal relationships with a human being. When considering a robot conforming to a humanlike appearance and learning behaviors, a psychological factor related to altruism and reciprocity naturally appears. This may be particularly important when considering teaching interactions. Creating a humanlike teaching scenario could reinforce the user's feeling that his or her role is to pass on knowledge to the robot, just as another human might help him or her out in a similar situation. In other words, the user would act as he or she would like to be treated.

The active teaching process presented in the proposed experiments probably evokes some of the feelings normally attributed to a caregiver-child dyad, in which the robot would elicit a certain form of attention. Indeed, watching the increased understanding of the skill by progressively lowering the scaffolds may be an important psychological factor for the teacher, where the user might feel self-esteem when teaching the less-knowledgeable robot. B. Lee et al. (2006) showed that behavioral synchronization is a powerful communicative experience for HRI. Through their experiments, the authors showed that in human dyadic interactions, close relationship partners were more engaged in joint attention to objects in the environments and to their respective body parts than strangers. In addition, they were more engaged in *discovering* and *showing* activities, sharing experience and exploring the environment together. Indeed, by contrasting a humanoid robot learning system with a computer language interpreting commands executed by a programmer, we see that both processes involve a transfer of information to the machine by incrementally watching the outcome of the transfer to help the user refine his or her teaching strategy. However, in the computer language situation, the considered machine is more of a tool to process the user's strategy rather than a peer learner that the user might wish to instruct. In robotics, Thomaz et al. (2005) presented HRI experiments in which the user could guide the robot's understanding of the objects in its environment. The interaction created a form of closeness where the user explored the environment together with the robot. In the experiments presented here, kinesthetic teaching also provided such a close relationship with the robot by physically guiding the robot's arms enabling it to collect kinesthetic experiences.

6.3.2 Insights from developmental sciences

Although animals such as chimpanzees can use tools to achieve a goal, they tend to discard the tools after using them. Humans however have developed the ability to recursively use tools (i.e., to use a tool to create another tool), attaching functions to objects that do not directly involve outcomes. This is believed to have led to the development of pedagogical skills as powerful social learning mechanisms, enabling the transmission of not just observable behaviors but also unobservable knowledge. Following this assumption, Gergely and Csibra (2005) presented experiments contrasting observational learning with pedagogy. Unlike observational learning, pedagogy requires an active participation of the teacher which is achieved by a special type of communication to manifest the relevant knowledge of a skill. The teacher does not simply use his or her knowledge, but engages in an activity that benefits the learner. To highlight which features of a skill are relevant, the teacher must first recognize the features by analyzing his or her knowledge relatively to the knowledge of the learner. Indeed, one does not need to be aware of knowledge content to generate an appropriate behavior. For example, riding a bike may appear easy for a person, but describing and teaching the skill to another person is not necessarily obvious.

Zukow-Goldring (2004) presented experiments in developmental psychology to understand the methods used by caregivers for assisting infants as they gradually learn new skills by engaging in new activities. By analyzing how a child learns manipulation and assembling skills (using pop beads), the authors showed that observation alone is insufficient. The caregiver first focuses the attention of the child to the *affordances* of the objects (the possibility to assemble the two objects). Because of the lack of information concerning the paths to follow for a correct assembly of the objects, the child first fails at reproducing the actions. The caregiver then shows what the child's body should do to assemble the objects with the required orientations and paths to follow, demonstrating the *effectivities* required for the assembly. The caregiver assists the child by partially demonstrating the action and by letting the child progressively resolve the skill by herself. Thus, the caregiver's gestures gradually provide perceptual information, guiding the infant to perform the skill. The task is first simplified and the child is then progressively put in various situations to experience and generalize the skill. *Embodying* and putting the child through the motions draws her attention to the coordination of *affordances* of the two objects as well as to the *effectivities* of the body required to connect the two objects. Thus, teaching is structured to permit the child to gather information on the characteristics of the objects and actions specific to each of them.

The experiments presented in this chapter have been inspired from this teaching process, where the proposed probabilistic framework naturally fits with the representation of *affordances* and *effectivities* learned simultaneously. As described in Zukow-Goldring (2004), by providing bodily experience, the caregiver provides the infant with the opportunity to see and feel the solutions to the correspondence problem (i.e., detecting the match between self and other). Similarly to the teaching process presented in these developmental psychology experiments, the proposed human-robot teaching scenario starts with the robot first observing the task performed by the user (through motion sensors). The robot learner then begins to reproduce the task through the teacher's assistance, gradually performing parts of the task independently (by a manual selection of the motors to control, *embodying* the robot and putting it through the motion). In these experiments, the *affordances* and *effectivities* refer respectively to the correct way of using the objects and to the correct movements that the robot's body must adopt in order to manipulate the objects appropriately.

Rohlfing et al. (2006) highlighted the importance of having multimodal cues to reduce the complexity of human-robot skill transfer. By considering multimodal information as an essential element for structuring the demonstrated tasks, they showed that humans transfer their knowledge through social interaction, by recognizing what current knowledge the learner lacks. Teachers are indeed sensitive to the cognitive abilities of their interaction partner, and taking this into account could reduce the learning complexity of current PbD frameworks. By considering that sharing human adaptability with the less knowledgeable is a central issue of social learning, it was hypothesized that a human teacher could also adapt naturally to a robot equipped with specific abilities. A similar strategy has been adopted in the experiments presented here by designing skill transfer processes that benefit from the user's capacity to adapt his or her teaching strategies to the particular context.

6.3.3   Insights from sociology

Vygotsky (1978) introduced the *zone of proximal development* (ZPD) as a general term to define the gap between what the learner already knows, namely, the learner's *zone of current development* (ZCD), and what can be acquired through the teacher's assistance (Wood & Wood, 1996). An efficient teaching strategy consists in exploring and being familiar with the ZPD of the learner in order to evaluate what the learner is able to acquire given his or her current ability. This general paradigm can also be applied to human-robot interaction, where the role of the teacher is to ascertain the current ZCD of the robot learner (by testing the robot's current understanding of the skill), and where its ZPD lies (i.e., to ascertain what can be achieved by the robot with assistance). Searching for what the robot already knows can appear time-consuming, but it may also constitute an important psychological factor for the teacher to help him/her feel involved in the collaborative human-robot teamwork.

When designing a human-robot teaching scenario, it is thus important to allow the teacher to acquire the robot's ZPD through interaction to provide individualized support to the robot. This is achieved by anticipating the problems that the robot might encounter, providing the appropriate scaffolds and gradually dismantling these scaffolds as the learner progresses (and eventually constructing further scaffolds for the next stage of learning). Similar to *molding* behaviors, a kinesthetic moving of the robot in its own environment provides a social way of assessing the robot's capacities and limitations when interacting with the environment.

6.3.4   Insights from sports science

To transfer a skill between two human partners, various means of performing demonstrations can be used depending on the motor skill that must be transferred. Several methodologies have been investigated for skill acquisition in sports with the aim of providing advice to sport coaches on how to efficiently transfer a motor skill and how to measure success depending on the capacities of individual athletes (Horn & Williams, 2004).

Coaching is the learning support aimed at improving the performance of the learner enabling him to carry out the task by providing directions and feedback; it is highly interactive and requires that one continuously analyzes the learner's accomplishment. To do so, the coach needs to be receptive to the learner's current performance level. A variety of modalities are traditionally used by sport coaches to help the athletes acquire the skill (Horn & Williams, 2004), where the visual observation of a skilled model completing the entire task provides a good basis for movement production. This also applies to other types of skills, as, for instance, training by expert surgeons (Custers, Regehr, McCulloch, Peniston, & Reznick, 1999). Similarly, in the experiments proposed in this chapter, the principal aim of observing the performance of an expert through motion sensors is to provide a complete and temporally continuous demonstration of the skill. By using motion sensors, the human expert can freely perform the task while the robot *observes* the full-body motion. However, as noted by Hodges and Franks (2002), optimal movement templates do not always generalize well across individuals. Thus, individually-based templates may be more appropriate in refining and achieving consistency in a skill. This supports the further use of kinesthetic learning to physically guide the robot's arms, permitting the robot to experience the skill.

This body of work also states that multiple exposures provide the opportunity to discern the structure of the modeled task. It allows the teacher to organize and verify what the learner knows and to focus attention on problematic aspects in subsequent exposures. This is achieved up to a *saturation point*, determined by the coach, at which additional demonstrations of the task yield no further learning benefits. Similarly, after each reproduction by the robot, the user decides whether the robot has correctly acquired the skill or whether further refinement is required. Several factors affect the amount of additional benefit that can be derived from multiple observations of a model. Among them is the isomorphism of the various demonstrations. Indeed, repetition of identical demonstrations may be of limited utility, whereas extremely diverse demonstrations may generate conflicts or confusion. To encourage flexibility and adaptability, the coaches often manipulate task properties (e.g., by using various situations) to provide appropriate variability depending on the learner's capacities. Similarly, in the proposed teaching scenario, the user is instructed to progressively displace the objects after each demonstration, thus providing variability in the exposures of the skill.

The different insights presented throughout this section originate from various fields of research but highlight the importance of a multimodal and incremental learning system to help the robot experience various situations and gradually refine its skill. This has been exploited in the proposed PbD framework by employing several modalities to demonstrate a skill and by using an incremental learning strategy to refine the model of the skill.

CHAPTER 7

USING SOCIAL CUES TO SPEED UP THE LEARNING PROCESS

This chapter presents an experiment where social cues are used as priors in the statistical learning framework in order to speed up the learning process.

For an efficient collaboration with human users, indoor robots such as humanoids should be provided with adaptive controllers that can behave robustly in changing situations. These robots should moreover be provided with natural interfaces to interact easily and naturally with end-users (Spexard, Hanheide, & Sagerer, 2007), and it should be possible to reprogram them in an intuitive manner (Billard, Calinon, Dillmann, & Schaal, 2008). Indeed, as these robots are supposed to use a very wide range of infrastructures and tools originally designed for humans, it is not possible to pre-encode all the gestures that will be required to perform manipulation-type skills. It is therefore crucial to facilitate the skill transfer process by providing end-users with natural teaching methods to reprogram these robots in an intuitive manner.

The statistical learning framework proposed in this book requires that the skill be observed in slightly different situations. Even if this variation appears naturally when executing the skill several times, we have seen in Chapter 6 that the robot's capacity to generalize over different contexts also depends on the pedagogical quality of the provided demonstrations (e.g., gradual variability of the situations and exaggerations of the key features to reproduce). This fact shares similarities with the human way of teaching. Indeed, a good teacher also extends the demonstrations progressively so that the learner can more easily infer the connections between the different examples, i.e., the range of situations where the skill may apply is progressively enlarged. Thus, by using solely statistical learning strategies, it is implicitly suggested that one way of increasing the speed of the teaching process is to rely on the user's natural propensity for teaching by structuring the successive examples provided and guiding the learner's exploration. For instance, in the stacking task scenario presented in Section 6.2.2, an expert user progressively displaced the cylinder and cube after each demonstration with the aim of providing variability in the exposures of the skill. In such a situation, it is nearly always possible for the robot to extract the task constraints with only a few demonstrations (from four to ten for most of the tasks considered in this book).

However, untrained users might provide a set of demonstrations remaining either too similar or too different from one example to the other. In this case, a larger set of demonstrations would be required to generalize the skill. To weaken the drawback of such situations, the learning approach is extended to the joint use of cross-situational observations and social cues to ensure an efficient transfer of the task through interaction with the user. We thus suggest here to enhance the statistical learning strategy with information originating from various social cues, and point out that these cues can be represented statistically as priors in the GMM/GMR framework. An experiment with gaze and speech information is presented to demonstrate that interactional cues of different nature can be considered.[1] Figure 7.1 presents the information flow across the complete system.

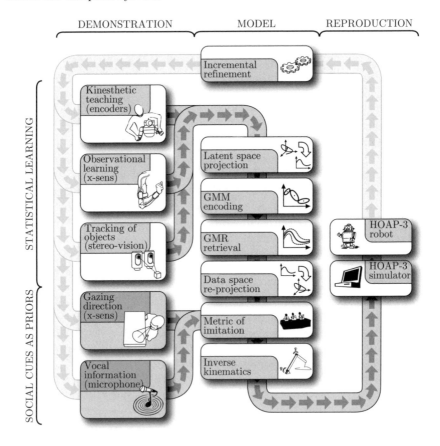

Figure 7.1 The information flow across the proposed system where various modalities are used to convey the demonstrations and where various social cues can be employed to scaffold the teaching interaction in order to speed up the learning process.

---

[1] This experiment does not aim at developing state-of-the-art gaze tracking systems or speech recognition engines. Prototypes of such systems are indeed proposed here to show that multimodal social cues can be integrated in a generic statistical learning approach.

By way of a computer game, Thomaz and Breazeal (2006) explored the ways in which machine learning can be designed to take advantage of natural human interaction and tutelage. They demonstrated that augmenting a reinforcement learning process with the social mechanisms of attention direction and gaze positively impacted the dynamics of the underlying learning mechanisms, highlighting the reciprocal nature of the teacher-learner partnership. They showed through experiments that the teacher's ability to guide the learner's attention to appropriate objects at appropriate timings creates a significantly more robust and efficient learning interaction. In the field of speech acquisition and word learning, Yu and Ballard (2007) investigated how humans can learn words/objects couplings through statistical learning, and proposed a model for early word acquisition in a unified framework integrating statistical and social cues. Their model links these two sources of information by considering joint attention and prosody on the one hand, and statistics from cross-situational observations on the other hand. Our work follows these two approaches by extending the concept to learning of continuous gestures in a humanoid robot. To speed up learning, a body of work was focused on the use of priors as part of the imitation learning process. Indeed, several hints can be used to transfer a skill, not only by demonstrating the task multiple times but also by highlighting the important components of the skill, which can be achieved by various means and through different modalities.

7.1 EXPERIMENTAL SETUP

7.1.1 Use of head/gaze information as priors

A large body of work explored the use of natural pointing and gazing cues as a means of conveying the intention of the user (Scassellati, 1999; Kozima & Yano, 2001; Ishiguro et al., 2003; Nickel & Stiefelhagen, 2003; Ito & Tani, 2004; Hafner & Kaplan, 2005; Breazeal et al., 2005; Kruijff, Zender, Jensfelt, & Christensen, 2007). In Calinon and Billard (2006), an investigation was carried out on how to extract gaze information by measuring the orientation of the head through *X-Sens* motion sensors. Even if this remains a strong assumption (as the head orientation cannot be directly considered to represent a social cue), it still affects gaze following, i.e., the head naturally turns towards a goal when there is no other constraint. Figure 7.2 illustrates the method used to detect joint attention by representing gaze direction as a cone of vision whose intersection with a table corresponds to a Gaussian distribution. Instead of simply defining an attention point as the intersection of a gaze direction line with a plane, this method also evaluates the robustness of the measure through a covariance matrix.

Gaze is here modeled by a cone of vision with vertex point t_1, direction d_1 and half-cone angle θ. A point x on the cone satisfies

$$d_1 \left(\frac{x - t_1}{|x - t_1|} \right) = \cos(\theta),$$

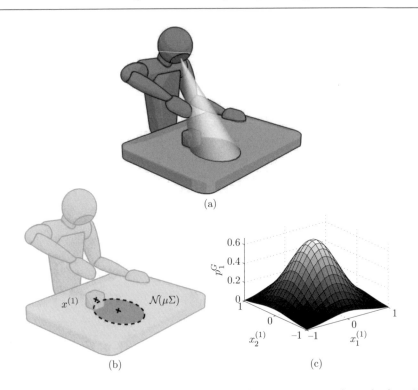

(a)

(b)   (c)

Figure 7.2 An illustration of the use of gaze information to speed up the learning process through a probabilistic measure of saliency when the head is turned toward an object. (a) The estimation of gaze direction by representing the position and orientation of the head (extracted through *X-Sens* motion sensors) as a cone of vision intersecting with a surface. (b) A probabilistic representation of the intersection as a 2D Gaussian distribution. (c) Estimation of the probability to focus the attention of the robot at a particular time on a particular object i placed on the table, given its initial position $o^{(i)}$ (the object position is tracked through the robot's built-in stereoscopic vision system).

or in a matrix form (where I denotes the identity matrix)

$$(x - t_1)^T M (x - t_1) = 0, \text{ with } M = d_1 d_1^T - (\cos(\theta))^2 I.$$

The table is defined by a plane with origin t_2, and directions d_{21} and d_{22}. A point x on the plane satisfies

$$x = t_2 + x_1 \, d_{21} + x_2 \, d_{22}.$$

The intersection of the cone and the plane defines a conic

$$c_1 \, x_1^2 + 2c_2 \, x_1 x_2 + c_3 \, x_2^2 + 2c_4 \, x_1 + 2c_5 \, x_2 + c_6 = 0,$$

with $t_{12} = t_2 - t_1$, $c_1 = d_{21}^T M d_{21}$, $c_2 = d_{21}^T M d_{22}$, $c_3 = d_{22}^T M d_{22}$, $c_4 = t_{12}^T M d_{21}$, $c_5 = t_{12}^T M d_{22}$ and $c_6 = t_{12}^T M t_{12}$.

This conic can be re-written in a homogenous matrix form as

$$x^T C x = 0, \quad C = \begin{pmatrix} c_1 & c_2 & c_4 \\ c_2 & c_3 & c_5 \\ c_4 & c_5 & c_6 \end{pmatrix} = \begin{pmatrix} C_R & C_t \\ C_t^T & C_\delta \end{pmatrix},$$

where $x = (x_1, x_2, 1)^T$, $C_R \in \mathbb{R}^{2 \times 2}$, $C_t \in \mathbb{R}^{1 \times 2}$ and $C_\delta \in \mathbb{R}$. The intersection of a cone and a plane forms an ellipse if

$$|C| \neq 0, \quad \left| \begin{pmatrix} c_1 & c_2 \\ c_2 & c_3 \end{pmatrix} \right| > 0, \quad \frac{|C|}{c_1 + c_3} < 0.$$

The canonical form of the conic C_c is determined by transforming the conic matrix C through an Euclidean transformation H

$$C_c = \begin{pmatrix} C_{c1} & 0 & 0 \\ 0 & C_{c2} & 0 \\ 0 & 0 & C_{c3} \end{pmatrix} = H^T C H, \quad \text{with} \quad H = \begin{pmatrix} R & t \\ 0^T & 1 \end{pmatrix},$$

where the rotation R and translation t are extracted by diagonalizing C_R through *Principal Component Analysis*

$$C_R = R \Lambda R^T, \qquad t = -R \Lambda^{-1} R^T C_t.$$

By considering an elliptical intersection (see also Fig. 7.2), the canonical conic $C_{c1} x_{c1}^2 + C_{c2} x_{c2}^2 + C_{c3} = 0$ can be re-written as

$$\frac{x_{c1}^2}{a^2} + \frac{x_{c2}^2}{b^2} = 1, \quad \text{with} \quad a = \sqrt{-\frac{C_{c3}}{C_{c1}}}, \quad b = \sqrt{-\frac{C_{c3}}{C_{c2}}},$$

which can also be represented as a 2D Gaussian distribution $\mathcal{N}(\mu, \Sigma) = \mathcal{N}(t, R \Sigma_c R^T)$, where Σ_c is defined by

$$\Sigma_c = \begin{pmatrix} a^2 & 0 \\ 0 & b^2 \end{pmatrix}.$$

By considering this Gaussian distribution in the proposed experiment, the likelihood $\mathcal{L}_{i,j}$ at time step t_j for an object i located at initial position $o^{(i)}$ can then be defined by

$$\mathcal{L}_{i,j} = \frac{1}{\sqrt{(2\pi)^2 |\Sigma_j|}} e^{-\frac{1}{2} \left((o^{(i)} - \mu_j)^T \Sigma_j^{-1} (o^{(i)} - \mu_j) \right)}.$$

When considering M objects, a probabilistic measure of interest (level of saliency) for an object i at time step t_j is similarly defined by

$$p_{i,j}^G = \frac{\mathcal{L}_{i,j}}{\sum_{n=1}^M \mathcal{L}_{n,j}} \in [0, 1]. \tag{7.1}$$

7.1.2 Use of vocal information as priors

Vocal deixis using speech recognition engines in a PbD context has been explored as a way for the user to highlight, through linguistic information, the steps of the demonstration that were deemed most important (Dominey et al., 2005; Dillmann, 2004; Pardowitz et al., 2007; Breazeal et al., 2005). Another approach has focused on the prosody of speech pattern rather than the linguistic content of speech, which was employed in a similar manner to infer information on the user's communicative intent (Rohlfing et al., 2006; Hegel, Spexard, Wrede, Horstmann, & Vogt, 2006; Breazeal & Aryananda, 2002; Breazeal et al., 2004; Thomaz et al., 2005). Certain approaches also combine both types of information, see e.g. Kruijff et al. (2007).

The prosodic approach is here followed by using *Hidden Markov Models* (HMMs) to detect particular intonation patterns while uttering attention to particular events during the demonstration of the task (e.g., emphasizing the use of a particular object). The method used for training and recognition of attentional utterances in the vocal trace using HMMs is presented next, showing how a vocal prior p^S can be retrieved from such a probabilistic model.

A set of 10 short common attentional utterances used as vocal spotlights produced by the user (such as *"Look here!"* or *"Watch this!"*) is first recorded in a training phase prior to the interaction. A series of 10 random words and/or sentences spoken in a neutral way (e.g., by reading an instruction) is also collected.

The pitch and energy of the sound signals are extracted, where the pitch (corresponding to the fundamental frequency f_0) is evaluated by the *subharmonic-to-harmonic ratio* method proposed by X. Sun (2002). The two sets of pitch and energy traces are used to train two *Hidden Markov Models* (HMMs) λ^A and λ^N ("attentional model" and "neutral model"), where the number of states has been determined empirically. Each HMM is thus defined by 3 states, where each observation output is defined by a 2D Gaussian distribution (to encode the pitch and energy). The parameters are trained through the *Baum-Welch algorithm* presented in Section 2.6.2, iteratively estimating the initial state distribution Π, the state transition probabilities a and the output distributions defined by centers μ and covariance matrices Σ.

Subsequently, when demonstrating a skill, the vocal trace of the user is recorded through the robot's internal microphone, which is used to detect the probability of an attentional bid, i.e., to estimate at which point in time the user brings the attention of the robot to a particular aspect of the skill during the course of his/her demonstration. To do so, a temporal window of fixed size W is used to keep track of the pitch and energy signals during the last W seconds (if $t_j < W$, then $W = t_j$). In this experiment, $W = 0.5\,\text{sec.}$ has been determined empirically. The data in this window are then tested with the two HMMs λ^A and λ^N at each time step t_j through the *forward procedure* presented in Section 2.3.1. $\mathcal{L}_j^A$ and $\mathcal{L}_j^N$ are thus computed, corresponding to the likelihoods at time t_j of belonging respectively to the "attentional model" λ^A or to the "neutral model" λ^N. At each time step t_j, the probability of detecting an attentional utterance $p_j^S \in [0,1]$ is thus defined by

$$p_j^S = \frac{\mathcal{L}_j^A}{\mathcal{L}_j^A + \mathcal{L}_j^N}.$$

(7.2)

Figure 7.3 presents an example of recording for the task of stacking a cylinder on a cube, where the first four rows represent results for the speech probability p^S extracted along the task, and where the last row corresponds to results for the gaze probability p^G, namely the probability of bringing the robot's attention to one of the objects detected in the scene (either the cylinder or the cube). The first row shows the sound signal corresponding to the sentence *"You take THIS and you put it THERE"* told by the user when executing the skill (see bottom-left snapshot in Fig. 6.1). We see that the particular events in the demonstration corresponding respectively to the subparts when the user

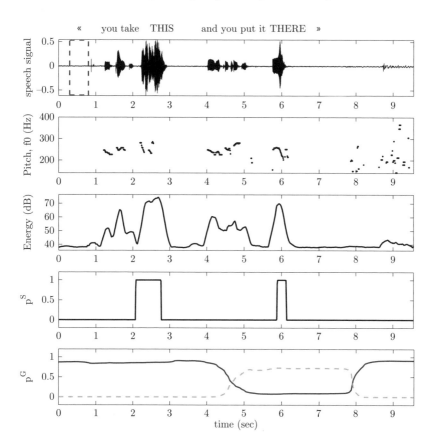

Figure 7.3 The extraction of priors from speech (*first 4 rows*) and gaze information (*last row*) for the stacking task presented in Section 6.2.2 (see Fig. 6.1 bottom). The first four graphs show the probabilistic extraction of attentional events in the vocal trace by using pitch and energy information (the temporal window of size W used to detect attentional cues is represented with a dashed line).

grasps one object (*"THIS"*) and drops it on the other object (*"THERE"*) are highlighted through the user's voice. These events correspond roughly to local patterns characterized by a higher energy and a larger pitch amplitude with consecutive rising and falling intonation contours, which are typical to prosodic patterns serving as spotlights during the interaction (Yu & Ballard, 2007), and which are automatically captured in the proposed system through HMM. The fourth graph represents the probability p_j^S at time t_j of hearing an attentional utterance. The last graph shows the probability p_j^G at time t_j of looking at the cylinder (*solid line*) or at the cube (*dashed line*), which also implicitly informs the robot that the user is conveying information on the relevance of these two objects at various time steps.

7.2   EXPERIMENTAL RESULTS

Figure 7.4 displays the results of an experiment similar to the one presented in Section 6.2.2 originally performed by an expert user. Here, an untrained user provides five demonstrations that are too similar for the task constraints to be correctly extracted only by statistical analysis.

We have seen in previous chapters that the statistical constraints relative to an object i can be represented by the generalized trajectory $\hat{\mu}^{(i)}$ and associated covariance matrices $\hat{\Sigma}^{(i)}$. By using this GMR representation, the influence of the constraints can be easily modified, as presented in Section 2.10, by taking into consideration at each time step t_j the gaze priors $p_{i,j}^G$ on object i as well as the speech priors p_j^S. To do so, the mean values $\bar{p}_{i,j}^G$ and $\bar{p}_j^S$ at time step t_j are computed by averaging over the various demonstrations provided to the robot, and a weighting factor scaling the covariance matrices of the GMR representation is defined such that

$$\hat{\Sigma}_j^{(i)'} = \hat{\Sigma}_j^{(i)}(1 - \gamma\, \bar{p}_{i,j}^G\, \bar{p}_j^S), \qquad (7.3)$$

where $\bar{p}_{i,j}^G\, \bar{p}_j^S$ represents the joint probability at time step t_j for object i, which serves as a spotlight to emphasize particular events during the demonstration. γ is a factor weighting the influence of the social cues over the constraints extracted through cross-situational statistics (here, $\gamma = 0.5$ has been selected empirically). Thus, for the reproduction of the task, the influence of the generalized trajectory with respect to object i is increased when $\bar{p}_{i,j}^G\, \bar{p}_j^S$ is high.

The imprecision due to the estimation of social cues is reduced by considering a variety of demonstrations and modalities. For example, in the bottom graph of Figure 7.3, we see that from time step $t_j = 8$ sec., the system detects that the user is looking at the initial position of the cylinder (which is already stacked on the cube). This error may be due to a tracking imprecision or be the result of the user no longer being required to focus on a particular object/position after having dropped the cylinder. In both cases, this error disappears by taking into consideration the joint probability of events (i.e., the vocal analysis does not detect a particular event at these time steps), as well as the multiple demonstrations provided to the robot.

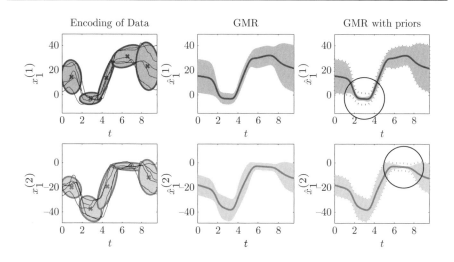

Figure 7.4 The influence of the speech/gaze priors on constraints extracted through statistical learning for an interaction with a naïve user. The first and second rows correspond to the trajectories related respectively to the cylinder and to the cube (see Fig. 6.1 bottom). *First column:* Data extracted from 5 demonstrations and encoding in a *Gaussian Mixture Model* (GMM) of 5 components. *Second column:* Generalization of the trajectories through *Gaussian Mixture Regression* (GMR) based solely on cross-situational statistics. In this case, we see that five demonstrations are insufficient for extracting interesting information concerning the task constraints, i.e., the envelope thickness around the generalized trajectory does not present much variation (see Fig. 6.6 for comparison). *Third column:* By using GMR with the social priors defined in Eqs. (7.1), (7.2) and (7.3), the envelopes become thinner in the relevant parts of the trajectories and for the relevant objects to consider, namely when grasping the cylinder and when dropping it on the cube (highlighted by two circles in the graphs).

The right column of Figure 7.4 shows the results of applying speech and gaze priors as proposed in Eq. (7.3) to the extraction of the task constraints (thinner envelope when $\bar{p}_{i,j}^{G}$ $\bar{p}_{j}^{S}$ is high). By using social cues as priors, the robot can thus generalize the skill to various situations, similarly to the reproduction results presented in Figure 6.6.

These results demonstrated that the integration of social cues within the proposed statistical learning approach was promising. However, as only a very limited dataset has been used so far, the robustness of the approach still needs to be evaluated with untrained users teaching new skills in real-world experimental setups. The weighting mechanism defined in Eq. (7.3) for combining prosodic and joint attention spotlights is rather simple and serves as a first step towards evaluating the integration of social cues within the statistical framework. A possible direction for future work would be to investigate the dependencies and relevance of these different cues in a human-robot teaching interaction context.

CHAPTER 8

DISCUSSION, FUTURE WORK AND CONCLUSIONS

This chapter discusses the proposed PbD framework, its advantages (Sect. 8.1) and limitations (Sect. 8.2), and suggests further research directions to pursue (Sect. 8.3).

8.1 ADVANTAGES OF THE PROPOSED APPROACH

8.1.1 Advantages of using motion sensors to track gestures

Modeling and generating a movement that appears similar to one produced by a human is no easy task. People can very easily distinguish a natural motion from one that is artificially generated but this is difficult to express in mathematical form, as is finding criteria to determine which motion can be considered as human-like. An approach proposed by Gams and Lenarcic (2006) involves taking into account a minimum torque in joints as a constraint, which implies having a dynamical model of the arm. Another approach is to apply statistical comparison with a large database of human poses. However, as the range of tasks that can be achieved with two arms is huge, and since humans move in different manners depending on their task and environment, such an approach usually requires the definition of task-specific criteria. Another strategy was followed in this book, which resided in using a reduced number of gestures captured by a human user to extract the key elements of a skill and by adapting the acquired skill to new situations. Thus, the system leveraged on human expertise to learn a task. As the user has optimized his/her motion skill throughout life, the perspective was taken that he/she could provide essential hints to a kinematically redundant robot for the reproduction of the skill, even if the reproduced movement cannot be exactly the same as the ones demonstrated due to dissimilar embodiments.

Observational learning in robotics is usually performed using vision systems since cameras represent the interfaces that are, technically, the most similar to human-like sensory systems. However, vision is sometimes incorrectly considered as being the unique choice of sensory system to be used for observation. In the experiments proposed throughout this book, motion sensors were often employed as an alternative to vision in order to *"observe"* the user's gestures. In Calinon and Billard (2006), we explored the use of motion sensors attached to the body of the demonstrator to convey information concerning human body gesture, and demonstrated that this method can present an alternative to vision systems for recording human motion data. The technique allowed the registering of natural movements with smooth velocity profiles characterizing human motion. The main advantage of motion sensors is that they usually provide more robust measures of the joint angles characterizing a human body posture. Nevertheless, the main advantage of vision as compared to motion sensors is that there is no need to wear special devices (Shon et al., 2005).[1] On the other hand, one drawback of vision concerns the fact that occlusion and lighting issues must be taken into account, which is usually not the case with motion sensors since they provide a continuous measure of the body gesture. If the vision system is mobile (e.g., embedded within a humanoid robot's head), the actuation of the head must also be taken into account by the tracking system. By using motion sensors, the measure of the joint angles between two segments can be easily extended to various body parts. This is often not the case for vision, which generally requires dedicated tracking algorithms depending on the body parts, i.e., legs and arms are often processed by different tracking algorithms (or tuned differently).

Another way of demonstrating a gesture is to utilize the kinesthetic teaching process presented in Section 3.2. Compared to the use of motion sensors attached to the body of the demonstrator, the joint angle trajectories recorded by kinesthetic teaching generally present sharper turns and an unnatural decorrelation of the various DOFs. Indeed, when physically grasping the end-effectors of the robot and displacing the joints in such a way, the motors of the arms tend to move sequentially in a decoupled manner. It was in Chapter 6 demonstrated through experiments that by combining information from the motion sensors and kinesthetic teaching, one could generate naturally-looking trajectories and at the same time tackle the correspondence problem. Indeed, due to the different embodiment between the user and the robot, it was not possible to directly copy the joint angle trajectories demonstrated through motion sensors. An efficient transfer of such a gesture often requires refinement of the trajectories with respect to the robot capabilities as well as to its environment. On the other hand, demonstrating a gesture only by kinesthetic teaching is limited by the naturalness of the motion and by the number of limbs that the user can control simultaneously. Combining both methods provides a social approach of teaching the humanoid robot. Thus, the use of motion sensors gave rise to a model of the entire movement for the robot, while kinesthetic teaching offered a way to refine the demonstrated movement by feeling the characteristics and limitations of the robot in its own environment.

---

[1] Note that a large set of vision systems also requires wearing special patches for a facilitated tracking by vision processing algorithms (Ude, Atkeson, & Riley, 2004; Shon et al., 2005).

8.1.2 Advantages of the HMM representation for imitation learning

In Calinon and Billard (2007c), we presented an implementation of an HMM-based system that was utilized prior to the development of the current GMM/GMR framework to encode, generalize, recognize and reproduce gestures by representing data in both visual and motor coordinates. In this work, we have described how various theoretical models of imitation learning have permitted the use of HMM in robotic applications, and how they fit the models of imitation proposed by other groups or in other fields of research. Indeed, some of the features contained in the HMM representation bear similarities with those of theoretical models of imitation. By using HMM to encode gestures, it is possible to maintain the theoretical aspects and architectures of these models, at the same time offering a probabilistic description of the data that is more suitable for a real-world robotic application. Here, we briefly summarize the considered models of imitation.

The *Associative Sequence Learning* (ASL) mechanism suggests that imitation requires a *vertical association* between a model's action, as viewed from the imitator's point of view, and the corresponding imitator's action (Heyes & Ray, 2000; Heyes, 2001). The vertical links between the sensory representation of the observed task and the motor representation are part of a repertoire, where elements can be added or refined. ASL suggests that the links are created essentially by experience, with a concurrent activation of sensory and motor representations. In their model, the mapping between the sensory and motor representation can be associated with a higher level representation (boxes depicted in Fig. 8.1). Indeed, the model assumes that data are correctly discretized, segmented and classified, similarly to the imitation model proposed by Byrne (1999). In the ASL model, the horizontal links simulate the successive activation of sensory inputs for learning the skill, activating simultaneously the corresponding motor representation to copy the observed behavior. Repetitive activation of the same sequence strengthens the links, aiding motor learning. Depending on the complexity of task organization, numerous demonstrations may be needed to provide sufficient data to extract the regularities. The more data that are available, the more evident is the underlying structure of the task,

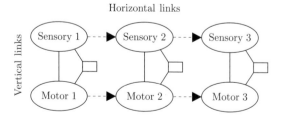

Figure 8.1 The Associative Sequence Learning (ASL) model inspired from Heyes (2001).

clarifying which elements are essential, which are optional, and which are variations in response to changing circumstances. Any action that the imitator is able to perform can also be recognized by observing a model demonstrating the task. This imitation model stresses the need for a probabilistic framework in a robotic application that can extract invariants across multiple demonstrations. We see that such a model is in agreement with an HMM decomposition of the task, where the underlying structure is learned by way of multiple demonstrations. If a given pattern appears frequently, its corresponding transition weights are automatically strengthened by the *Baum-Welch* learning algorithm presented in Section 2.6.2.

Each hidden state in an HMM can output multimodal data. In Calinon and Billard (2007c), we used this property to model multiple variables within various frames of reference. Thus, we encoded a skill by using motion sensors to record joint angle data (motor representation) and hand paths tracked by a stereoscopic vision system (visual representation). In this application, the role of HMM was to create a link between the datasets, where each state can be considered as a label (or as a higher level representation) common to the visual and motor data. Indeed, in HMM, the sequence of states is not observed directly and generates the visual or motor representation. Thus, similarly to the ASL mechanism, the architecture of a multivariate HMM also has a horizontal process in order to associate the elements in a sequential order (by learning transition probabilities between the hidden states), as well as a vertical process to associate each sensory representation to appropriate motor representation (which is done through the hidden state). If data are missing from one part or the other (visual or motor representation), it is still possible to recognize a task, and retrieve a generalization of the task in the other representation space.

Nehaniv and Dautenhahn (2002, 2001) also suggested the use of a general algebraic framework to address the correspondence problem in both natural and artificial systems. In this framework, they introduced the notion of a *correspondence problem* to refer to the problem of creating an appropriate mapping between what is performed by the demonstrator and what is reproduced by the imitator, where the correspondences can be constructed at various levels of granularity, reflecting the choice of a sequence of subgoals. Indeed, the two agents considered as a demonstrator and an imitator may not share the same embodiments (e.g., they may present differences in limb lengths, sensors or actuators). In their framework, the authors interpret the ASL model by representing the behavior performed by the demonstrator and imitator as two automata structures, with states and transitions. The *states* are the basic elements segmented from the whole task, that can produce *effects* or responses in the environment (e.g., a displaced object), and an *action* is the transition from one state to another (Fig. 8.2). An imitation process is then defined as a partial mapping process between the demonstrator's and the imitator's *states*, *effects*, and *actions*. In their model, an observer (i.e., an external observer, the demonstrator or the imitator) decides which of the *states*, *actions* or *effects* that are the most important to imitate by determining a metric of imitation. A variety of metrics are then used to yield solutions to multiple correspondence

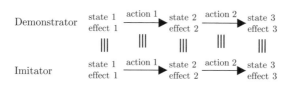

Figure 8.2 An algebraic framework inspired by Nehaniv and Dautenhahn (2002) to map the states, effects and actions of the demonstrator and the imitator.

problems, thus also rendering it possible to formally quantify the success of an imitation.

This representation is closely related to the characteristics of HMM. Indeed, HMM offers a direct extension of the automata depicted in this algebraic framework, where the main advantage lies in its stochastic representation of the transitions and observations, which is more suitable for encoding real-world data such as the ones collected by the robot in a PbD framework. To draw links with their algebraic model, the hidden states of an HMM output probabilistically distributed values that are similar to the notion of *effects*, while transitions between hidden states (also probabilistically defined) are equivalent to *actions*. Note that the encoding of data in HMM does not resolve the correspondence problem but provides a suitable representation to treat the *what-to-imitate* and the *how-to-imitate* issues in a common framework. An HMM framework thus offers a stochastic method to model the process underlying gesture imitation, by making a link between theoretical concepts and practical applications. In particular, it stresses the fact that the observed elements of a demonstration and the organization of these elements should be stochastically described to have a robust robot application taking into account the high variability and the discrepancies across both demonstrator and imitator points of view.

8.1.3   Advantages of the GMR representation for regression

Section 2.4 presented the regression method that was used to retrieve smooth generalized versions of the demonstrated trajectories based on Gaussian Mixture Models encoding. We discuss here the similarities shared by *Gaussian Mixture Regression* (GMR) with other regression frameworks proposed in the literature.

In *Projection Pursuit Regression* (PPR), high-dimensional data are iteratively interpreted in lower-dimensional projections where a smoothing method is applied (Friedman, Jacobson, & Stuetzle, 1981). In *Multivariate Adaptive Regression Splines* (MARS), high-dimensional data are split into subspaces where splines are fitted to the data (Friedman, 1991). Xu, Jordan, and Hinton (1995) proposed an approach to modeling data with a mixture of experts trained by EM algorithm, where the experts take the form of polynomial functions (linear models in that case), resulting in a piecewise polynomial approximation of the data. The use of local regression techniques is probably the most extensively studied technique, initially aimed at smoothing data. We refer to Cleveland and Loader (1996) for a review. By considering a dataset of input and output variables $\{x_i, y_i\}_{i=1}^{N}$, the objective of local regression is to estimate $\mathbb{E}(y_i) = f(x_i)$ locally, by considering a subset of points $\{x_j\}$ in the neighborhood of x_i.

Cleveland (1979) introduced *Locally Weighted Regression* (LWR) where the fit also depends on the distance between x_j and x_i. The influence of x_j is thus weighted, giving more weight to the points near x_i whose response is being estimated, and less weight to the points further away. For each point x_i whose response is being estimated, a linear or quadratic polynomial function is usually applied to fit the data locally. By computing the regression function values for each of the N datapoints, a smooth curve is then retrieved where the smoothness is determined by the weighting function. When considering a Gaussian distribution centered on x_i as the weighting function,[2] the amount of data contributing to the local estimation can be parameterized by the variance Σ_i.[3] The selection of Σ_i is thus a bias-variance trade-off (Geman, Bienenstock, & Doursat, 1992), i.e., a large Σ_i produces a smooth estimate (small variance), while a small Σ_i conforms more to the data (small bias). Similarly, the choice of the polynomial degree is also a bias-variance tradeoff. Methods have been proposed to automatically select these parameters, namely an adaptive local bandwidth selection and an adaptive local selection of the polynomial degree (Cleveland & Loader, 1996). As LWR combines the simplicity of linear least squares regression and the flexibility of nonlinear regression, the approach soon appeared very appealing for robot control (Atkeson, 1990; Atkeson et al., 1997; Moore, 1992). As discussed in Section 1.4, work based on this LWR approach was mainly concentrated on moving on from a memory-based approach to a model-based technique, and moving on from a batch learning process to an incremental learning strategy.

Schaal and Atkeson (1998) introduced *Receptive Field Weighted Regression* (RFWR) as a non-parametric approach to incrementally learn the fitting function without having to store the whole training data (i.e., without using historical data). The method is based on a receptive field approach where each receptive field has the form of a Gaussian kernel that is updated independently. The process renders it possible to add and prune receptive fields and to deal with irrelevant inputs. Within each receptive field, a local linear function models the relationships between input data X and output data Y. The strategy resembles Gaussian mixture modeling, where each Gaussian would be trained separately. Indeed, by encoding $\{X, Y\}$ as a joint distribution in a GMM of parameters $\{\mu_k, \Sigma_k\}_{k=1}^{K}$, the local Gaussian model of X is described by $\{\mu_k^X, \Sigma_k^X\}$, which is closely related to the notion of a Gaussian unit in the

---

[2] Usually, the distribution is computed only on a bounded interval for computational efficiency.

[3] To draw a link with the generalization approach proposed in this book, LWR would be defined as the estimate of a response $\hat{\xi}^{\mathcal{O}}$ for a given input $\xi^{\mathcal{I}}$, computed by weighting the output variables $\xi^{\mathcal{O}}$ with respect to their closeness to the input variable $\xi^{\mathcal{I}}$.

receptive field. In receptive fields, the parameters are $\{c_k, M_k\}_{k=1}^K$, representing respectively the centers (that are not updated) and distance metric of the Gaussian (that are similar to Σ_k^X). The local Gaussian model of Y is parameterized by $\{\mu_k^Y, \Sigma_k^Y\}$, which can also be related to the local linear model in the receptive field. In receptive fields, the parameters of the linear model are denoted $\{\beta_k\}_{k=1}^K$ and are learned either incrementally or in batch mode. In a batch mode, a leave-one-out cross validation scheme can be used to learn $\{M_k\}_{k=1}^K$ by a gradient descent method with a penalty factor regulating the smoothness. An incremental version can then be derived by a stochastic approximation. Section 2.6.3 demonstrated that when considering GMM, an incremental version of the well-known *Expectation-Maximization* (EM) algorithm can be used to update the parameters when new data are available. The iterative estimation process can also be regulated by bounding the covariance matrices (see Sect. 2.9.1), a strategy very similar to using a penalty factor to regulate the smoothness in the RFWR gradient descent method.

To resolve the *curse of dimensionality* issue (Scott, 2004) in RFWR, Vijayakumar et al. (2005) suggested improving the approach in order for it to operate efficiently in high-dimensional space through *Locally Weighted Projection Regression* (LWPR). Indeed, for high-dimensional data, it is far from optimal to consider the distance between a set of points in a receptive field approach due to the distance among points not causing a well enough separation in high dimensions. By detecting locally redundant or irrelevant input dimensions, the approach suggests to locally reduce the dimensionality of the input data. The curse of dimensionality can, for example, be avoided by finding local projections of the input data through *Partial Least Squares* (PLS) regression (Wold, 1966). Projection regression decomposes multivariate regressions into a superposition of univariate regressions along the projections in the input space. Multiple methods have been proposed to determine which are the efficient projections (Vijayakumar & Schaal, 2000), and PLS was found to be the most suitable technique for reducing the dimensionality in such a regression framework.

The proposed framework follows a similar strategy by considering that a set of trajectories can be represented locally by Gaussian distributions. Principal Component Analysis is first used as a global projection method to reduce, if possible, the dimensionality of the dataset, and to decorrelate the dataset globally. Subsequently, by using GMM, the non-linearities of the trajectories are modeled by piecewise local information in the form of Gaussian distributions locally representing the variability and the correlations of the data. GMM can be viewed as something in-between a local and global model, in the sense that it locally fits the joint input and output data by way of a global learning algorithm. GMR is used as a generic retrieval process to estimate generalized multivariate output signals that take the form of multivariate Gaussian distributions given multivariate input signals.

The principal advantage of using GMM/GMR for the considered applications is that GMM can be readily employed to recognize trajectories and estimate reproduction attempts, while GMR utilizes the same representation to extract constraints and retrieve a generalized version of the trajectories. The other advantage consists in the associated *Expectation-Maximization* learning methods being very easy to implement, thus providing a simple estimation approach known to have decent convergence properties if appropriately used. As GMM is a popular encoding scheme, the approach can also benefit from advances in various research domains and can be easily integrated in other probabilistic framework.

In essence, *Gaussian Process Regression* (GPR) shares similarities with the GMM/GMR approach proposed in this book (Rasmussen & Williams, 2006). It has been successfully applied to robotics to generalize over a set of demonstrations, see e.g. Grimes et al. (2006). Similarly to Support Vector Regression (SVR) (Smola & Schölkopf, 2004), GPR however suffers from high computational complexity which prevents its usage for large numbers of samples or online learning to date. A local variant of Gaussian processes has recently been proposed in robotics to deal with this issue (Nguyen-Tuong & Peters, 2008). Most regression models including GPR, SVR and LWPR are discriminative in that they learn $\mathcal{P}(Y|X)$ without modeling the joint distribution $\mathcal{P}(X,Y)$. The point of view of this book was to use an alternative generative approach by first solving the most general problem of estimating $\mathcal{P}(X,Y)$. Based on Gaussian Mixture Regression, it was assumed that $\mathcal{P}(X,Y)$ can be modeled by a finite set of Gaussian distributions. Regression on this model results in a mixture of linear models and naturally extends to regression with multiple outputs. Retrieving multiple outputs and their associated correlations is, on the contrary, not straightforward for GPR or SVR, which requires the development of other strategies to deal with multiple outputs (Boyle & Frean, 2005).

Even if these approaches share many similarities, it remains advantageous to utilize GMR for the applications considered here since: (1) it allows to deal very flexibly with recognition and reproduction issues in a common probabilistic framework; and (2) the learning process is fast and distinct from the retrieval process. Thus, a simple learning process can be used to model the demonstrated skill during the phases of the interaction that do not require real-time computation (i.e., after the demonstrations), and where a faster regression process is used afterwards for controlling the robot in an online manner during the reproduction phases. Thus, the multivariate input variables and multivariate output variables can be specified on-the-fly during reproduction as the joint distribution has been previously learned by the model, thus rendering the approach robust to various types of missing variables.

8.2   FAILURES AND LIMITATIONS OF THE PROPOSED APPROACH

8.2.1   Loss of important information through PCA

In the proposed PbD framework, we suggested to use PCA as a pre-processing step to reduce the dimensionality of the data and to decorrelate them by projection in a latent space of motion (see Sect. 2.7). We showed in Sect. 3.2

that when considering a small dataset of high dimensionality, this reduction of the dimensionality was a sufficient pre-processing step for further encoding in GMM. When recording data of high dimensions, the noise inherent to the demonstrations and the errors due to the tracking system are often encapsulated in the dataset. Reducing the dimensionality thus becomes a good opportunity to denoise the signals. However, some interesting features of the signals can also be lost during the projection. For example, by considering the motion of a hand while wiping a blackboard, the recorded gesture can be decomposed into a small circular motion (aiming at wiping the local part of the blackboard) and a large motion (aiming at covering the whole surface of the blackboard). In this example, PCA could fail at extracting the small circular motion since PCA retains the dimensions presenting a large variability. Here, PCA would thus fail at efficiently representing the purpose of the task, and Independent Component Analysis would provide a better performance. Similarly, when considering PCA as a criterion to reduce the dimensionality and by recording gestures of a person while chopping parsley with a knife, the processing of the data in a latent space of motion of lower dimensionality also fails at extracting the rapid mincing gesture of the hand. Figure 8.3 presents such an example where the joint angle θ_8 corresponds to an important oscillatory signal for the task. Due to its small variance, this variable is not taken into account when reducing the dimensionality of the latent space of motion. We see that this essential characteristic of the task (which in this case is not due to noise) disappears when reproducing the skill (*bold line*).

8.2.2 Failures at learning incrementally the GMM parameters

Throughout the experiments presented in this book, the user has demonstrated skills in a variety of situations to help the robot extract the task constraints and generalize the skill to various contexts. However, the situations encountered should not be too dissimilar from one demonstration to the following one, i.e., the skill should be demonstrated with an increased variability in the situations where it can be applied. This issue shares similarities with the teaching paradigm in humans. When a teacher presents examples of a skill, these examples must be sufficiently connected to show how the skill can be extended to different situations. In other words, the consecutive demonstrations should show a progression in the generalization perspective.

In the proposed system, if the user shows consecutive examples that are very much disconnected from one another, the learning system can fail at appropriately connecting the examples and at converging to an optimal solution. This failure can be partially prevented by updating the model only with trajectories that it has previously recognized, i.e., if the likelihood of the model is above a given threshold. Thus, the model is updated only if the demonstration provided is not too dissimilar to the situations already encountered. Figure 8.4 illustrates this issue, where two models are built incrementally upon the same set of demonstrations but presented in a different order. The models are learned through the *direct-update method* presented in Section 2.6.3. When the skill

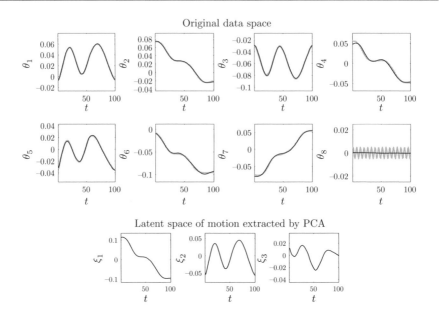

Figure 8.3 A failed attempt at reproducing a skill when reducing the dimensionality through a PCA pre-processing phase that projects the dataset in a latent space of lower dimensionality. *Top:* Original joint angle data (*grey line*) recorded when demonstrating a *"chopping parsley"* motion with a knife on a cutting board. The last joint angle represents the small oscillation of the wrist used to mince the parsley. *Bottom:* The 3-dimensional latent space of motion extracted by PCA. The dataset in this latent space is then projected back in the original data space (*black line*). We see that by pre-processing the highly dimensional data through PCA, the important oscillatory information contained in θ_8 is lost during re-projection of the data (i.e., the oscillatory motion disappears).

is progressively demonstrated (i.e., when two consecutive examples share similarities), encoding and generalization of the skill provide very satisfying results (*top*). However, when two consecutive demonstrations are presented in very different situations, i.e., when an attempt has been made to generalize the skill too rapidly, the local solution found by the system is less optimal (*bottom*). We see indeed that after the second demonstration, the system is unable to combine the last part of the first trajectory with the last part of the second trajectory (only a poor local solution is found by the incremental version of EM). The two final parts of the first and second trajectories thus tend to be encoded separately, as observed after the fifth demonstration. One Gaussian is stuck to the encoding of only a specific part of the dataset (the end of the first demonstration). In other words, the component is not updated optimally with respect to the other demonstrations. This is not very surprising since the assumption of the direct update algorithm is not satisfied; at least for the first two demonstrations. Indeed, to produce a correct update of the parameters, the set of posterior probabilities should remain similar enough when new data

Consecutive demonstrations that share similarities

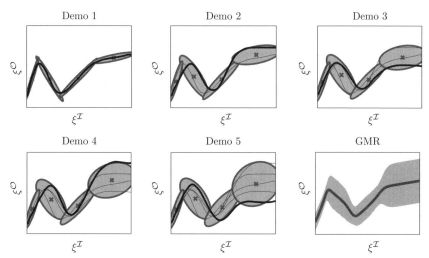

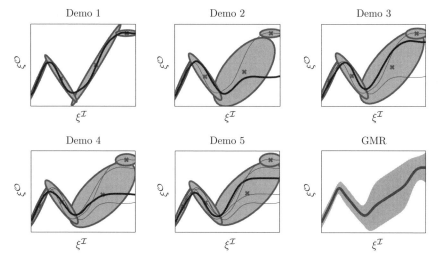

Figure 8.4 An illustration of the importance of the variability between two consecutive demonstrations when incrementally learning gestures using the *direct update method* presented in Section 2.6.3. The same data are used in the two situations but are not presented to the model in the same order. The graphs show the consecutive encoding of the data in GMM after convergence of the EM algorithm. Only the latest observed trajectory (*bold line*) is used to update the models (no historical data is used). The graphs in the last column show the corresponding generalization through GMR.

are used to update the model. Thus, the update of the GMM parameters provides a poor estimate after the second demonstration, and the model remains stuck with this poor local optimum in the next iterations.

Figure 8.5 presents a similar example concerning the incremental learning of the data using the *generative method* presented in Section 2.6.3. For this particular dataset, we see that the generative method is better at capturing the

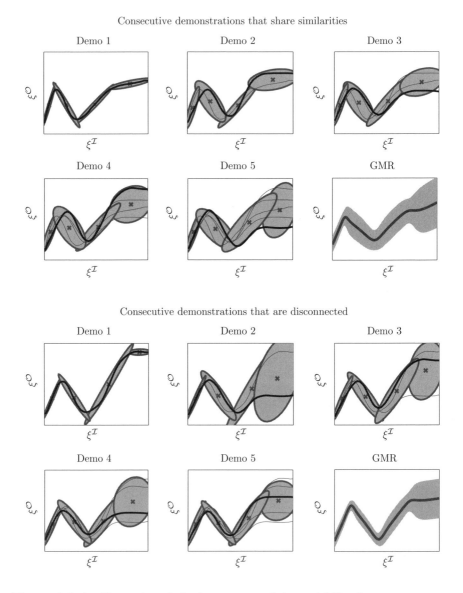

Figure 8.5 An illustration of the importance of the variability between two consecutive demonstrations when incrementally learning the data using the *generative method* (see also Fig. 8.4).

essential constraints of the task in terms of variability. However, relying on a stochastic process still has the disadvantage that it cannot guarantee that two consecutive learning processes provide the same estimation of the parameters.

8.2.3   Failures at extending a skill to a context that is too dissimilar to the ones encountered

After having statistically extracted the task constraints, we have seen in Section 2.4 how regression can be used to reproduce the skill in new situations that have not been observed during the demonstrations. This is true for situations that are similar to the ones demonstrated, but the generalization capability remains limited to the variability of the provided demonstrations. Figure 8.6 presents an example of failure for the *Bucket Task* presented in Section 4.3. We see in graph ① that the essential characteristics of the task are correctly reproduced, namely grasping the bucket and bringing it to a specific place. However, in graphs ② and ③, where the initial positions of the bucket are far from the demonstrated trajectories, the system fails at correctly reproducing

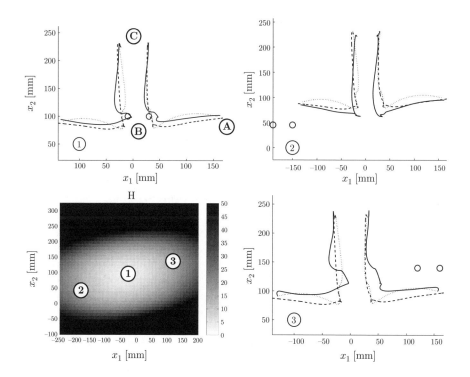

Figure 8.6 A failed attempt at extending the task to a broader context. *Bottom left:* A map depicting values of the cost function $\mathcal{H}$ for the *"Chess Task"* with respect to the initial position of the chess piece on the board. The graphs ①, ② and ③ present the reproduction attempts for the corresponding locations on the map (see legend in Fig. 4.14).

the skill (only a small deviation of the trajectories in the direction of the bucket is observed). Thus, in a situation that is too dissimilar to the ones encountered during the demonstrations, the system may fail at reproducing the skill due to the robot's body limitation and to the restrictive generalization capabilities of the learning system which depends on the observed variability.

8.3   FURTHER ISSUES

8.3.1   Toward combining exploration and imitation

One should recall that a main argument for the development of PbD methods was that it would speed up learning by providing examples of "good solution". However, as was seen in Section 8.2.3, this was true insofar as the context for the reproduction was sufficiently similar to that of the demonstration. In order to allow the robot to learn how to extend a task to any new situation, it appears important to combine PbD methods with other motor learning techniques (Billard et al., 2008).

The proposed statistical learning framework system was successfully extended by Guenter, Hersch, Calinon, and Billard (2007) to re-use a learned skill in a new context by letting the robot explore novel solutions through *Reinforcement Learning* (RL) techniques. Figure 8.7 illustrates the principles of this joint work, and the search process is presented in Figure 8.8. The GMM representations learned during the demonstrations are employed as hints for the search of new solutions during the self-exploration phase. The exploration of novel solutions is initiated by first displacing the centers μ_k of the GMM (with respect to Σ_k) and by using GMR to retrieve new trajectories that are smooth enough to be considered as plausible solutions for the robot. This process renders it possible to extend the skill to a larger context than the ones encountered during the demonstrations. Exploration is indeed complementary to imitation, and the proposed application highlights the generality of using a probabilistic framework to represent the task constraints.

Within the framework of this book, we have focused on learning a skill by imitation. It is however known in human development that there are other means of acquiring new skills. Self-experimentation and practice are used jointly with imitation, where the observations initiate further search of a controller by repeated reproduction attempts. This combination of observation and practice aims at developing more robust and autonomous controllers leveraging from the use of new demonstrations when the robot is faced with a situation that is too dissimilar to the ones previously demonstrated (Atkeson & Schaal, 1997; Bentivegna et al., 2004).

Reinforcement learning (RL) appears particularly suited for this type of problem. Early work on PbD using RL began in the 1990s and featured the learning of how to control an inverse pendulum and make it swing up (Atkeson et al., 1997). More recent efforts have focused on the robust control of the upper body of humanoid robots while performing various manipulation tasks (Peters et al., 2003; Yoshikai et al., 2004; Bentivegna et al., 2004). A body of

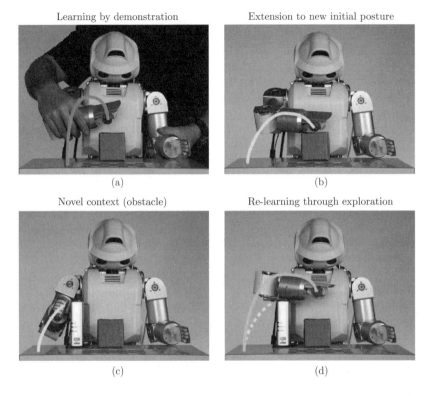

Figure 8.7 The joint use of imitation and exploration for acquiring a skill. (a) The robot is trained through kinesthetic demonstrations on a task that consists in placing an object in a box. (b) By imitation, the robot is successful in reproducing the skill when the new situation differs only slightly from that of the demonstration. (c) The robot fails at reproducing the skill when the context changes considerably (a large obstacle has been placed in the way). (d) The robot relearns a new trajectory that reproduces the essential aspect of the demonstrated skill, namely reaching the box by avoiding the obstacle and dropping the object into it. Adapted from Guenter et al. (2007).

work has also explored this issue within an evolutionary perspective by creating populations of agents that copy each other in order to learn a control strategy by conducting experiments themselves and by watching others (Billard & Dautenhahn, 1998; Hwang, Choi, & Hong, 2006; Jansen & Belpaeme, 2006).

8.3.2   Toward a joint use of discrete and continuous constraints

Throughout this book, the constraints of a skill have been represented in a continuous form, based on the perspective that a continuous model is generic and allows the encoding of various kinds of movements by carrying the degree

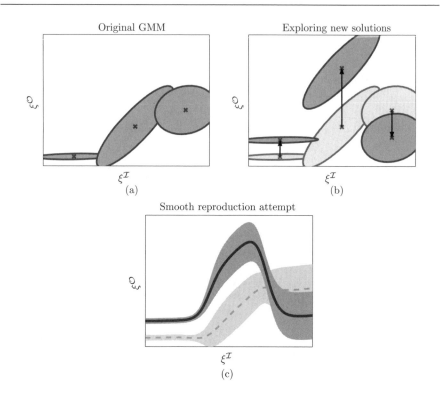

Figure 8.8 The use of self-experimentation to explore the search space for new solutions based on the examples originally provided by the user. (a) The original model of the skill learned through multiple demonstrations. (b) The search of new solutions by randomly displacing the centers of the 3 Gaussians with respect to the learned covariance information and by using *Reinforcement Learning* to find a correct adaptation of the skill. (c) A new smooth trajectory generated through GMR by using the modified model (the original model is represented in a lighter shade of the color).

of invariance along the trajectories. In future work, a segmentation of these constraints could also be considered in order to detect particular events in the flow of motion. Indeed, as seen throughout the experiments, parts of the tasks are characterized by particular events that may be represented in a discrete form (such as grasping or dropping an object). It would thus be important in future work to consider such constraints as higher-level events by segmenting the trajectories into a discrete set of important features, i.e., by segmenting the continuous flow of motion with respect to the subparts demonstrating strong invariance.

The BMM representation of binary signals presented in Section 2.11.1 could, for example, be used as a hint to segment the whole motion into specific events. Indeed, BMM is able to probabilistically model the occurrence of particular events in the trajectory (e.g., when a keyword is detected by a speech recognition system), and the probabilistic representation of BMM renders it possible to efficiently handle the variability observed across multiple demonstrations. Discretization would thus allow a motion to be represented as a set of discrete events that are invariant across multiple demonstrations, and for which the transitions from one event to the other are described in a continuous form using GMM/GMR. For reproduction, the advantage of decomposing a motion into a set of separate subtasks is that each subtask can be refined separately, e.g., by producing a demonstration only for one of the subtasks instead of for the whole motion. Moreover, if the robot fails at reproducing one subpart of the task, it would have the possibility to try again to reproduce the failed subpart and then continue with the remaining parts without having to start over from the beginning of the motion. By using this segmentation process, the various subtasks could also be combined in different orders by taking advantage of the smoothing properties of GMR.

The automatic extraction of specific events in the trajectory is closely related to the *when-to-imitate* issue and to the automatic segmentation of human motion data (Kulic et al., 2008). In the experiments presented in this book, the user manually partitioned the data trace into segments that were used in further processing. In Calinon and Billard (2004), we tested a simple method to automatically segment multiple demonstrations based on zero velocity detection. We noticed through this experiment that the process was robust when used by trained users but that it required adaptation from users interacting with the robot for the first time. This pointed at the unnaturalness of the process. The *when-to-imitate* issue thus needs to be further explored to detect particular events in the continuous flow of motion.

If the whole motion is segmented as a sequence of subtasks, the elements can be organized in a different order to perform new skills. One way to accomplish this is to extend the proposed framework to hierarchical models. By taking inspiration from the issues in PbD depicted in Muench et al. (1994), the generalization issue can be considered both at a *trajectory level* and at a *process level* through a hierarchical architecture. Indeed, GMMs can be used to encode basic trajectory elements represented as states in a *Dynamic Bayesian Network* (DBN) or in a *Hidden Markov Model*. The role of the DBN/HMM would be to learn the skill in terms of organization of trajectory elements. A generalized sequence of elements can then be retrieved by the model. After organization these basic elements, GMR can be used to smoothly connect the various motion segments, which would produce a low-level controller for the robot characterized by smooth transitions between motion primitives. The generalization would thus take place simultaneously at a *trajectory level* to extract the relevant features of an action, and at a *program level* to organize the different components of a behavior (Byrne, 1999).

Another crucial point to consider in future work is the *who-to-imitate* issue, or more generically how to automatically select good examples for the learning of a task. Indeed, the current system does not cope with outliers and the learning system currently relies on the quality of the examples provided by the user. This is currently left as an open question and further investigations should explore how to give the robot the ability to select who is an appropriate teacher for a given situation, and how to cope with poor demonstrations (Ting, D'Souza, & Schaal, 2007).

8.3.3   Toward predicting the outcome of a demonstration

Another direction of research concerns the ability of the robot to understand the intention of the user and the ability to learn a skill by observing a failed attempt from the demonstrator. Even by watching a failed attempt, a human learner can infer the desired result that the teacher intended to achieve. It would be interesting to provide the robot with such an ability, which is also related to the dual role of perception and prediction explored in robotics (Y. Demiris & Hayes, 2002). Y. Demiris and Khadhouri (2006) suggested employing building blocks consisting of pairs of inverse and forward models that compete in parallels and in a hierarchical manner to both plan and execute an action. This architecture allows treating both perception and actions in a unified model, and renders it possible to provide an estimate of the upcoming outcomes when observing an action performed by the demonstrator. An hypothetic solution to deal with such an issue in the proposed probabilistic framework would be to take advantage of the prediction properties of the GMM. Indeed, after observing a set of successful attempts, the model has the ability to recognize a skill on the fly by computing the likelihood of the model when faced with the partial observation of the skill performed by the user. This on-line recognition property would then make it possible to estimate the purpose of a task (or at least to estimate which are the following particular task constraints of the motion), assuming that correct demonstrations of the skill have previously been demonstrated. Prior information in the form of attentional scaffolding would also be of high relevance to extract the intent of the user more robustly, as discussed in Chapter 7.

Further issues also concern the plasticity of the learning process. Indeed, Section 2.6.3 presented two incremental learning algorithms that equally weighed the various demonstrations presented to the robot. However, this linear learning curve remains a strong assumption that does not match the learning rate of humans. Considering the notion of plasticity in the proposed learning framework (or alternatively considering forgetting factors) would allow to fix motions in a repertoire as soon as the skill has been observed and practiced a sufficient number of times. This would indeed become a crucial element for lifelong robot learning (Thrun & Mitchell, 1995). The early experiences of a motion could also be toned down by forgetting the oldest examples and focusing exclusively on the recent demonstrations, which would allow the robot to adapt its behavior to slowly changing constraints. A direction for future work would be to explore this trade-off and define learning policies to modulate how the learning system should use its past and current experience in an optimal manner.

8.4   FINAL WORDS

Throughout this book, we explored the use and combination of various machine learning techniques in a *Robot Programming by Demonstration* (PbD) framework to probabilistically encode a task represented as a trace of sensory variables collected by the robot. We showed that encapsulating the local properties of these multivariate signals in a probabilistic model made it possible to deal with recognition, generalization, reproduction and evaluation issues in a unified PbD framework. We demonstrated how a robot could autonomously extract task constraints in a continuous and probabilistic form to automatically determine an optimal controller to reproduce the skill robustly in different situations and in case of perturbations.

We presented how a generic model based on *Gaussian Mixture Model* (GMM) and *Gaussian Mixture Regression* (GMR) could be incrementally trained to continuously represent the constraints of a skill and to retrieve a robust controller for the robot. The various algorithms were presented didactically and supplied with simple examples. Sourcecodes for these examples were also made available online in *Matlab* and C/C++ to help the reader test, modify and reproduce the proposed experiments. Several regularization and pre-processing methods were also proposed for an efficient use in a PbD context. We then emphasized the use of these probabilistic methods in *Human-Robot Interaction* (HRI) and proposed active teaching methods placing the human teacher in the loop of the robot's learning by incrementally acquiring a skill through scaffolding methods and by utilizing different modalities to demonstrate gestures.

The generality of the approach was demonstrated through various experiments and collaborative work showing that the proposed framework can be used co-jointly with other approaches. We first presented experiments to test and optimize the algorithms proposed within a PbD context, and then presented various human-robot teaching experiments to didactically illustrate the use of the proposed algorithms by highlighting both technical and social learning perspectives.

Robot programming by demonstration is a growing area in robotics, and we see the use of probabilistic models in this field as a rich and principled framework to endow robots with highly robust controllers permitting them to deal with uncertainty and noisy sensors to behave in real-world situations. Statistical learning representation also renders it possible to establish generic connections between different models in robotics from learning, perception, motor control or planning perspectives, as well as to provide an opportunity to enable existing models from various disciplines to coexist (e.g., by integrating social cues to motor control). We are eagerly looking forward to further studies in this field (both in theory and practice) and are confident that many important and appealing developments will strive toward this direction in the near future.

For updated information and sourcecodes, visit the book's companion website, http://www.programming-by-demonstration.org/book.

REFERENCES

Aleotti, J., & Caselli, S. (2005, August). Trajectory clustering and stochastic approximation for robot programming by demonstration. In *Proc. IEEE/RSJ intl conf. on intelligent robots and systems (IROS)* (pp. 1029–1034).

Aleotti, J., & Caselli, S. (2006a, May). Grasp recognition in virtual reality for robot pregrasp planning by demonstration. In *Proc. IEEE intl conf. on robotics and automation (ICRA)* (pp. 2801–2806).

Aleotti, J., & Caselli, S. (2006b). Robust trajectory learning and approximation for robot programming by demonstration. *Robotics and Autonomous Systems, 54*(5), 409–413.

Aleotti, J., Caselli, S., & Reggiani, M. (2004). Leveraging on a virtual environment for robot programming by demonstration. *Robotics and Autonomous Systems, 47*(2-3), 153–161.

Alissandrakis, A., Nehaniv, C., & Dautenhahn, K. (2002). Imitation with ALICE: Learning to imitate corresponding actions across dissimilar embodiments. *IEEE Trans. on Systems, Man, and Cybernetics, Part A: Systems and Humans, 32*(4), 482–496.

Alissandrakis, A., Nehaniv, C., & Dautenhahn, K. (2007). Correspondence mapping induced state and action metrics for robotic imitation. *IEEE Trans. on Systems, Man and Cybernetics, Part B, 37*(2), 299–307.

Alissandrakis, A., Nehaniv, C., Dautenhahn, K., & Saunders, J. (2005). An approach for programming robots by demonstration: Generalization across different initial configurations of manipulated objects. In *Proc. IEEE intl symposium on computational intelligence in robotics and automation* (pp. 61–66).

Alissandrakis, A., Nehaniv, C., Dautenhahn, K., & Saunders, J. (2006, March). Evaluation of robot imitation attempts: Comparison of the system's and the human's perspectives. In *Proc. ACM SIGCHI/SIGART conf. on human-robot interaction (HRI)* (pp. 134–141).

Arandjelović, O., & Cipolla, R. (2006). Incremental learning of temporally-coherent Gaussian mixture models. *Technical Papers - Society of Manufacturing Engineers (SME)*.

Arbib, M., Billard, A., Iacoboni, M., & Oztop, E. (2000). Mirror neurons, imitation and (synthetic) brain imaging. *Neural Networks, 13*(8-9), 953–973.

Arbib, M., Billard, A., Iacobonic, M., & Oztop, E. (2000). Special issue on synthetic brain imaging: Grasping, mirror neurons and imitation. *Neural Networks*, *13*, 975–997.

Arbib, M., Iberall, T., & Lyons, D. (1985). Coordinated control program for movements of the hand. *Experimental brain research Suppl.*, *10*, 111–129.

Ardizzone, E., Chella, A., & Pirrone, R. (2000, July). Pose classification using support vector machines. In *Proc. IEEE-INNS-ENNS intl joint conf. on neural networks (IJCNN)* (Vol. 6, pp. 317–322).

Asfour, T., Gyarfas, F., Azad, P., & Dillmann, R. (2006, December). Imitation learning of dual-arm manipulation tasks in humanoid robots. In *Proc. IEEE-RAS intl conf. on humanoid robots (Humanoids)* (pp. 40–47).

Atkeson, C. (1990). Using local models to control movement. In *Advances in neural information processing systems (NIPS)* (Vol. 2, pp. 316–323).

Atkeson, C., Hale, J., Pollick, F., Riley, M., Kotosaka, S., Schaal, S., et al. (2000). Using humanoid robots to study human behavior. *IEEE Intelligent Systems*, *15*(4), 46–56.

Atkeson, C., Moore, A., & Schaal, S. (1997). Locally weighted learning for control. *Artificial Intelligence Review*, *11*(1-5), 75–113.

Atkeson, C., & Schaal, S. (1997). Learning tasks from a single demonstration. In *Proc. IEEE/RSJ intl conf. on robotics and automation (ICRA)* (Vol. 2, pp. 1706–1712).

Azuma, R., Baillot, Y., Behringer, R., Feiner, S., Julier, S., & MacIntyre, B. (2001). Recent advances in augmented reality. *IEEE Computer Graphics and Applications*, *21*(6), 34–47.

Bakker, P., & Kuniyoshi, Y. (1996, April). Robot see, robot do : An overview of robot imitation. In *Proc. workshop on learning in robots and animals (AISB)* (pp. 3–11).

Bejczy, A., Kim, W., & Venema, S. (1990, May). The phantom robot: predictive displays for teleoperation with time delay. In *Proc. IEEE intl conf. on robotics and automation (ICRA)* (p. 546–551).

Bekkering, H., Wohlschlaeger, A., & Gattis, M. (2000). Imitation of gestures in children is goal-directed. *Quarterly Journal of Experimental Psychology*, *53A*(1), 153–164.

Belpaeme, T., Boer, B. de, & Jansen, B. (2007). The dynamic emergence of categories through imitation. In K. Dautenhahn & C. Nehaniv (Eds.), *Imitation and social learning in robots, humans and animals: Behavioural, social and communicative dimensions*. Cambridge, UK: Cambridge University Press.

Bentivegna, D., Atkeson, C., & Cheng, G. (2004). Learning tasks from observation and practice. *Robotics and Autonomous Systems*, *47*(2-3), 163–169.

Billard, A. (2002). Imitation: a means to enhance learning of a synthetic proto-language in an autonomous robot. In K. Dautenhahn & C. Nehaniv (Eds.), *Imitation in animals and artifacs* (pp. 281–311). Cambridge, MA, USA: MIT Press.

Billard, A. (2003). Robota: Clever toy and educational tool. *Robotics and Autonomous Systems*, *42*, 259–269.

Billard, A., Calinon, S., Dillmann, R., & Schaal, S. (2008). Robot programming by demonstration. In B. Siciliano & O. Khatib (Eds.), *Handbook of robotics* (pp. 1371–1394). Secaucus, NJ, USA: Springer.

Billard, A., Calinon, S., & Guenter, F. (2006). Discriminative and adaptive imitation in uni-manual and bi-manual tasks. *Robotics and Autonomous Systems*, *54*(5), 370–384.

Billard, A., & Dautenhahn, K. (1998). Grounding communication in autonomous robots: an experimental study. *Robotics and Autonomous Systems*, *24*(1-2), 71–79.

Billard, A., & Dillmann, R. (2006). Special issue on the social mechanisms of robot programming by demonstration. *Robotics and Autonomous Systems*, *54*(5), 351–352.

Billard, A., & Hayes, G. (1999). Drama, a connectionist architecture for control and learning in autonomous robots. *Adaptive Behavior*, *7*(1), 35–64.

Billard, A., & Mataric, M. (2001). Learning human arm movements by imitation: Evaluation of a biologically-inspired connectionist architecture. *Robotics and Autonomous Systems*, *37*(2), 145–160.

Billard, A., & Schaal, S. (2006). Special issue on the brain mechanisms of imitation learning. *Neural Networks*, *19*(3).

Bilmes, J. (1998). *A gentle tutorial on the EM algorithm and its application to parameter estimation for Gaussian mixture and hidden Markov models.* Technical Report, University of Berkeley, ICSI-TR-97-021.

Bishop, C. (2006). *Pattern recognition and machine learning (information science and statistics).* Secaucus, NJ, USA: Springer.

Borga, M. (1998). *Learning multidimensional signal processing.* PhD thesis, Linköping University, Sweden.

Boyle, P., & Frean, M. (2005). Dependent gaussian processes. In *Advances in neural information processing systems 17* (pp. 217–224). MIT Press.

Bradski, G. (1998, October). Real time face and object tracking as a component of a perceptual user interface. In *Proc. IEEE workshop on applications of computer vision (WACV)* (pp. 214–219).

Brand, M., & Hertzmann, A. (2000, July). Style machines. In *Proc. ACM intl conf. on computer graphics and interactive techniques (SIGGRAPH)* (pp. 183–192).

Brass, M., Bekkering, H., Wohlschlaeger, A., & Prinz, W. (2001). Compatibility between observed and executed movements: Comparing symbolic, spatial and imitative cues. *Brain and Cognition*, *44*(2), 124–143.

Breazeal, C., & Aryananda, L. (2002). Recognition of affective communicative intent in robot-directed speech. *Autonomous Robots*, *12*(1), 83–104.

Breazeal, C., Berlin, M., Brooks, A., Gray, J., & Thomaz, A. (2006). Using perspective taking to learn from ambiguous demonstrations. *Robotics and Autonomous Systems*, *54*(5), 385–393.

Breazeal, C., Brooks, A., Gray, J., Hoffman, G., Kidd, C., Lee, H., et al. (2004). Tutelage and collaboration for humanoid robots. *Humanoid Robots*, *1*(2), 315–348.

Breazeal, C., Buchsbaum, D., Gray, J., Gatenby, D., & Blumberg, B. (2005). Learning from and about others: Towards using imitation to bootstrap the social understanding of others by robots. *Artificial Life*, *11*(1-2), 31–62.

Brethes, L., Menezes, P., Lerasle, F., & Hayet, J. (2004, April). Face tracking and hand gesture recognition for human-robot interaction. In *Proc. IEEE intl conf. on robotics and automation (ICRA)* (Vol. 2, pp. 1901–1906).

Bullock, D., & Grossberg, S. (1988). Neural dynamics of planned arm movements: Emergent invariants and speed-accuracy properties during trajectory formation. *Psychological Review*, *95*(1), 49–90.

Bullock, D., & Grossberg, S. (1989). VITE and FLETE: Neural modules for trajectory formation and postural control. In W. Hersberger (Ed.), *Volitional control* (pp. 253–297). Elsevier.

Burdea, G. (1999). Invited review: the synergy between virtual reality and robotics. *IEEE Trans. on Robotics and Automation*, *15*(3), 400–410.

Butcher, J. (2003). *Numerical methods for ordinary differential equations.* Chichester, UK: Wiley.

Byrne, R. (1999). Imitation without intentionality: Using string parsing to copy the organization of behaviour. *Animal Cognition*, *2*, 63–72.

Calinon, S., & Billard, A. (2004, September). Stochastic gesture production and recognition model for a humanoid robot. In *Proc. IEEE/RSJ intl conf. on intelligent robots and systems (IROS)* (pp. 2769–2774).

Calinon, S., & Billard, A. (2005, August). Recognition and reproduction of gestures using a probabilistic framework combining PCA, ICA and HMM. In *Proc. intl conf. on machine learning (ICML)* (pp. 105–112).

Calinon, S., & Billard, A. (2006, September). Teaching a humanoid robot to recognize and reproduce social cues. In *Proc. IEEE intl symposium on robot and human interactive communication (RO-MAN)* (pp. 346–351).

Calinon, S., & Billard, A. (2007a, August). Active teaching in robot programming by demonstration. In *Proc. IEEE intl symposium on robot and human interactive communication (RO-MAN)* (pp. 702–707). (Best paper award)

Calinon, S., & Billard, A. (2007b, March). Incremental learning of gestures by imitation in a humanoid robot. In *Proc. ACM/IEEE intl conf. on Human-Robot Interaction (HRI)* (pp. 255–262).

Calinon, S., & Billard, A. (2007c). Learning of gestures by imitation in a humanoid robot. In K. Dautenhahn & C. Nehaniv (Eds.), *Imitation and social learning in robots, humans and animals: Behavioural, social and communicative dimensions* (pp. 153–177). Cambridge, UK: Cambridge University Press.

Calinon, S., & Billard, A. (2007d). What is the teacher's role in robot programming by demonstration? - Toward benchmarks for improved learning. *Interaction Studies*, *8*(3), 441–464.

Calinon, S., Guenter, F., & Billard, A. (2005, April). Goal-directed imitation in a humanoid robot. In *Proc. IEEE intl conf. on robotics and automation (ICRA)* (pp. 299–304).

Calinon, S., Guenter, F., & Billard, A. (2007). On learning, representing and generalizing a task in a humanoid robot. *IEEE Trans. on Systems, Man and Cybernetics, Part B*, *37*(2), 286–298.

Cannon, D., & Thomas, G. (1997). Virtual tools for supervisory and collaborative control of robots. *Presence: Teleoperators and Virtual Environments*, *6*(1), 1–28.

Carreira-Perpiñán, M. (2002). *Continuous latent variable models for dimensionality reduction and sequential data reconstruction*. PhD thesis, Dept. of Computer Science, University of Sheffield, UK.

Caurin, G., Albuquerque, A., & Mirandola, A. (2004, September-October). Manipulation strategy for an anthropomorphic robotic hand. In *Proc. IEEE/RSJ intl conf. on intelligent robots and systems (IROS)* (pp. 1656–1661).

Chalodhorn, R., Grimes, D., Maganis, G., Rao, R., & Asada, M. (2006, May). Learning humanoid motion dynamics through sensory-motor mapping in reduced dimensional spaces. In *Proc. IEEE intl conf. on robotics and automation (ICRA)* (pp. 3693–3698).

Chan, T., & Dubey, R. (1995). A weighted least-norm solution based scheme for avoiding joints limits for redundant manipulators. *IEEE Trans. on Robotic and Automation*, *11*, 286–292.

Chella, A., Dindo, H., & Infantino, I. (2006). A cognitive framework for imitation learning. *Robotics and Autonomous Systems*, *54*(5), 403–408.

Chiaverini, S., Oriolo, G., & Walker, I. (2008). Kinematically redundant manipulators. In B. Siciliano & O. Khatib (Eds.), *Handbook of robotics* (pp. 245–268). Springer.

Chiu, C., Chao, S., Wu, M., & Yang, S. (2004). Content-based retrieval for human motion data. *Visual Communication and Image Representation*, *15*, 446–466.

Cleveland, W. (1979). Robust locally weighted regression and smoothing scatterplots. *American Statistical Association*, *74*(368), 829–836.

Cleveland, W., & Loader, C. (1996). Smoothing by local regression: Principles and methods. In W. Haerdle & M. G. Schimek (Eds.), *Statistical theory and computational aspects of smoothing* (pp. 10–49). New York: Springer.

Cohn, D., Ghahramani, Z., & Jordan, M. (1996). Active learning with statistical models. *Artificial Intelligence Research*, *4*, 129–145.

Cuijpers, R., van Schie, H., Koppen, M., Erlhagen, W., & Bekkering, H. (2006). Goals and means in action observation: A computational approach. *Neural Networks*, *19*(3), 311–322.

Custers, E., Regehr, G., McCulloch, W., Peniston, C., & Reznick, R. (1999). The effects of modeling on learning a simple surgical procedure: See one, do one or see many, do one? *Advances in Health Sciences Education*, *4*(2), 123–143.

Cypher, A., Halbert, D., Kurlander, D., Lieberman, H., Maulsby, D., Myers, B., et al. (1993). *Watch what I do: Programming by demonstration*. Cambridge, MA, USA: The MIT Press.

Dasgupta, S., & Schulman, L. (2000). A two-round variant of EM for Gaussian mixtures. In *Proc. conf. on uncertainty in artificial intelligence (UAI)* (pp. 152–159).

Dautenhahn, K. (1995). Getting to know each other - Artificial social intelligence for autonomous robots. *Robotics and Autonomous Systems*, *16*(2-4), 333–356.

Decety, J., Chaminade, T., Grezes, J., & Meltzoff, A. (2002). A PET exploration of the neural mechanisms involved in reciprocal imitation. *Neuroimage*, *15*(1), 265–272.

Delson, N., & West, H. (1996). Robot programming by human demonstration: Adaptation and inconsistency in constrained motion. In *Proc. IEEE intl conf. on robotics and automation (ICRA)* (pp. 30–36).

Demiris, J., & Hayes, G. (1996). Imitative learning mechanisms in robots and humans. In *Proc. European workshop on learning robots* (pp. 9–16).

Demiris, Y., & Hayes, G. (2002). Imitation as a dual route process featuring predictive and learning components: a biologically-plausible computational model. In K. Dautenhahn & C. Nehaniv (Eds.), *Imitation in animals and artifacs* (pp. 327–361). Cambridge, MA, USA: MIT Press.

Demiris, Y., & Khadhouri, B. (2006). Hierarchical attentive multiple models for execution and recognition of actions. *Robotics and Autonomous Systems*, *54*(5), 361–369.

Dempster, A., & Rubin, N. L. D. (1977). Maximum likelihood from incomplete data via the EM algorithm. *Journal of the Royal Statistical Society B*, *39*(1), 1–38.

Dillmann, R. (2004). Teaching and learning of robot tasks via observation of human performance. *Robotics and Autonomous Systems*, *47*(2-3), 109–116.

Dillmann, R., Kaiser, M., & Ude, A. (1995, July). Acquisition of elementary robot skills from human demonstration. In *Proc. intl symposium on intelligent robotic systems (SIRS)* (pp. 1–38).

Dominey, P., Alvarez, M., Gao, B., Jeambrun, M., Cheylus, A., Weitzenfeld, A., et al. (2005, December). Robot command, interrogation and teaching via social interaction. In *Proc. IEEE-RAS intl conf. on humanoid robots (Humanoids)* (pp. 475–480).

Dufay, B., & Latombe, J.-C. (1984). An approach to automatic robot programming based on inductive learning. *Robotics Research*, *3*(4), 3–20.

Ehrenmann, M., Rogalla, O., Zoellner, R., & Dillmann, R. (2001). Teaching service robots complex tasks: Programming by demonstation for workshop and household environments. In *Proc. IEEE intl conf. on field and service robotics (FRS)*.

Ekvall, S., & Kragic, D. (2005, April). Grasp recognition for programming by demonstration. In *Proc. IEEE intl conf. on robotics and automation (ICRA)* (pp. 748–753).

Ekvall, S., & Kragic, D. (2006, September). Learning task models from multiple human demonstrations. In *Proc. IEEE intl symposium on robot and human interactive communication (RO-MAN)* (pp. 358–363).

Erlhagen, W., Mukovskiy, A., Bicho, E., Panin, G., Kiss, C., Knoll, A., et al. (2006). Goal-directed imitation for robots: A bio-inspired approach to action understanding and skill learning. *Robotics and Autonomous Systems*, *54*(5), 353–360.

Fagg, A., & Arbib, M. (1998). Modeling parietal-premotor interactions in primate control of grasping. *Neural Networks*, *11*(7), 1277–1303.

FIBA Central Board. (2006). *The official basketball rules*. Intl Basket Federation (FIBA).

Fodor, I. (2002). *A survey of dimension reduction techniques*. Technical Report, Lawrence Livermore National Laboratory, Center for Applied Scientific Computing, UCRL-ID-148494.

Friedman, J. (1991). Multivariate adaptive regression splines. *The Annals of Statistics*, *19*(1), 1–67.

Friedman, J., Jacobson, M., & Stuetzle, W. (1981). Projection pursuit regression. *American Statistical Association*, *76*, 817–823.

Friedrich, H., Muench, S., Dillmann, R., Bocionek, S., & Sassin, M. (1996). Robot programming by demonstration (RPD): Supporting the induction by human interaction. *Machine Learning*, *23*(2), 163–189.

Fritsch, F., & Carlson, R. (1980). Monotone piecewise cubic interpolation. *SIAM Journal on Numerical Analysis*, *17*(2), 238–246.

Fujie, S., Ejiri, Y., Nakajima, K., Matsusaka, Y., & Kobayashi, T. (2004, September). A conversation robot using head gesture recognition as para-linguistic information. In *IEEE intl workshop on robot and human interactive communication (RO-MAN)* (pp. 159–164).

Furui, S. (1986). Speaker-independent isolated word recognition using dynamic features of speech spectrum. *IEEE Trans. on Acoustics, Speech, and Signal Processing*, *34*(1), 52–59.

Gams, A., & Lenarcic, J. (2006). Humanoid arm kinematic modeling and trajectory generation. In *IEEE/RAS-EMBS intl conf. on biomedical robotics and biomechatronics (BioRob)* (pp. 301–305).

Gaussier, P., Moga, S., Banquet, J., & Quoy, M. (1998). From perception-action loop to imitation processes: a bottom-up approach of learning by imitation. *Applied Artificial Intelligence*, *7*(1), 701–729.

Geman, S., Bienenstock, E., & Doursat, R. (1992). Neural networks and the bias/variance dilemma. *Neural Computation*, *4*(1), 1–58.

Gergely, G., & Csibra, G. (2005). The social construction of the cultural mind: Imitative learning as a mechanism of human pedagogy. *Interaction Studies*, *6*, 463–481.

Ghahramani, Z., & Jordan, M. (1994). Supervised learning from incomplete data via an EM approach. In J. D. Cowan, G. Tesauro, & J. Alspector (Eds.), *Advances in neural information processing systems* (Vol. 6, pp. 120–127). Morgan Kaufmann Publishers, Inc.

Grimes, D., Chalodhorn, R., & Rao, R. (2006, August). Dynamic imitation in a humanoid robot through nonparametric probabilistic inference. In *Proc. robotics: Science and systems (RSS)*.

Grochow, K., Martin, S., Hertzmann, A., & Popovic, Z. (2004). Style-based inverse kinematics. In *Proc. ACM intl conf. on computer graphics and interactive techniques (SIGGRAPH)* (pp. 522–531).

Gross, R., & Shi, J. (2001, June). *The CMU motion of body (MoBo) database* (Tech. Rep. No. CMU-RI-TR-01-18). Pittsburgh, PA: Robotics Institute, Carnegie Mellon University.

Guenter, F., Hersch, M., Calinon, S., & Billard, A. (2007). Reinforcement learning for imitating constrained reaching movements. *Advanced Robotics*, *21*(13), 1521–1544.

Gupta, A., & O'Malley, M. (2006, June). Design of a haptic arm exoskeleton for training and rehabilitation. *IEEE/ASME Trans. on Mechatronics*, *11*(3), 280–289.

Hafner, V., & Kaplan, F. (2005). Learning to interpret pointing gestures: Experiments with four-legged autonomous robots. In S. Wermter, G. Palm, & M. Elshaw (Eds.), *Biomimetic neural learning for intelligent robots* (Vol. 3575, p. 225–234). Springer.

Hegel, F., Spexard, T., Wrede, B., Horstmann, G., & Vogt, T. (2006, December). Playing a different imitation game: Interaction with an empathic android robot. In *Proc. IEEE-RAS intl conf. on humanoid robots (Humanoids)* (pp. 56–61).

Heinzmann, J., & Zelinsky, A. (1999). A safe-control paradigm for humanrobot interaction. *Intelligent and Robotic Systems*, *25*(4), 295–310.

Heise, R. (1989). *Demonstration instead of programming: Focussing attention in robot task acquisition*. PhD thesis, University of Calgary, Alberta Canada.

Hersch, M. (2008). *Adaptive sensorimotor peripersonal space representation and motor learning for a humanoid robot*. PhD thesis, Ecole Polytechnique Fédérale de Lausanne (EPFL), Switzerland.

Hersch, M., Guenter, F., Calinon, S., & Billard, A. (2008). Dynamical system modulation for robot learning via kinesthetic demonstrations. *IEEE Trans. on Robotics*, *24*(6), 1463–1467.

Heyes, C. (2001). Causes and consequences of imitation. *Trends in Cognitive Sciences*, *5*, 253–261.

Heyes, C., & Ray, E. (2000). What is the significance of imitation in animals? *Advances in the Study of Behavior*, *29*, 215–245.

Hightower, J., & Borriello, G. (2001, August). Location systems for ubiquitous computing. *IEEE Computer*, *34*(8), 57–66.

Hill, P. (1973). *Al-Jazari: The book of knowledge of ingenious mechanical devices*. Dordrecht, The Netherlands: Springer.

Hodges, N., & Franks, I. (2002). Modelling coaching practice: the role of instruction and demonstration. *Sports Sciences*, *20*(10), 793–811.

Hoffman, M., Grimes, D., Shon, A., & Rao, R. (2006). A probabilistic model of gaze imitation and shared attention. *Neural Networks*, *19*(3), 299–310.

Hollerbach, J., & Jacobsen, S. (1996, October). Anthropomorphic robots and human interactions. In *Proc. intl symposium on humanoid robots (HURO)* (pp. 83–91).

Horn, R., & Williams, A. (2004). Observational learning: Is it time we took another look? In A. Williams & N. Hodges (Eds.), *Skill acquisition in sport: Research, theory and practice* (pp. 175–206). London, UK: Routledge.

Hovland, G., Sikka, P., & McCarragher, B. (1996). Skill acquisition from human demonstration using a hidden Markov model. In *Proc. IEEE intl conf. on robotics and automation (ICRA)* (pp. 2706–2711).

Hwang, Y., Choi, K., & Hong, D. (2006, May). Self-learning control of cooperative motion for a humanoid robot. In *Proc. IEEE intl conf. on robotics and automation (ICRA)* (pp. 475–480).

Hyvärinen, A. (1999). Fast and robust fixed-point algorithms for independent component analysis. *IEEE Trans. on Neural Networks*, *10*(3), 626–634.

Iacoboni, M., Woods, R., Brass, M., Bekkering, H., Mazziotta, J., & Rizzolatti, G. (1999). Cortical mechanisms of human imitation. *Science*, *286*, 2526–2528.

Iba, S., Weghe, J., Paredis, C., & Khosla, P. (1999). An architecture for gesture-based control of mobile robots. In *Proc. IEEE/RSJ intl conf. on intelligent robots and systems (IROS)* (Vol. 2, pp. 851–857).

Ijspeert, A., Nakanishi, J., & Schaal, S. (2002a). Movement imitation with nonlinear dynamical systems in humanoid robots. In *Proc. IEEE intl conf. on robotics and automation (ICRA)* (pp. 1398–1403).

Ijspeert, A., Nakanishi, J., & Schaal, S. (2002b). Movement imitation with nonlinear dynamical systems in humanoid robots. In *Proc. IEEE intl conf. on robotics and automation (ICRA)* (pp. 1398–1403).

Ijspeert, A., Nakanishi, J., & Schaal, S. (2003). Learning control policies for movement imitation and movement recognition. In *Neural information processing system (NIPS)* (Vol. 15, pp. 1547–1554).

Inamura, T., Kojo, N., & Inaba, M. (2006, October). Situation recognition and behavior induction based on geometric symbol representation of multimodal sensorimotor patterns. In *Proc. IEEE/RSJ intl conf. on intelligent robots and systems (IROS)* (pp. 5147–5152).

Inamura, T., Kojo, N., Sonoda, T., Sakamoto, K., Okada, K., & Inaba, M. (2005). Intent imitation using wearable motion capturing system with online teaching of task attention. In *Proc. IEEE-RAS intl conf. on humanoid robots (Humanoids)* (pp. 469–474).

Inamura, T., Toshima, I., & Nakamura, Y. (2003). Acquiring motion elements for bidirectional computation of motion recognition and generation. In B. Siciliano & P. Dario (Eds.), *Experimental robotics VIII* (Vol. 5, pp. 372–381). Springer-Verlag.

International Federation of Robotics (IFR). (2008). *World robotics 2008: Statistics, market analysis, forecasts, case studies and profitability of robot investment*.

Ishiguro, H., Ono, T., Imai, M., & Kanda, T. (2003). Development of an interactive humanoid robot robovie - An interdisciplinary approach. *Robotics Research*, *6*, 179–191.

Ito, M., Noda, K., Hoshino, Y., & Tani, J. (2006). Dynamic and interactive generation of object handling behaviors by a small humanoid robot using a dynamic neural network model. *Neural Networks*, *19*(3), 323–337.

Ito, M., & Tani, J. (2004). Joint attention between a humanoid robot and users in imitation game. In *Proc. intl conf. on development and learning (ICDL)*.

Jansen, B., & Belpaeme, T. (2006). A computational model of intention reading in imitation. *Robotics and Autonomous Systems*, *54*(5), 394–402.

Jenkins, O., & Mataric, M. (2004). A spatio-temporal extension to isomap nonlinear dimension reduction. In *Proc. intl conf. on machine learning (ICML)* (pp. 56–63).

Jolliffe, I. (2002). *Principal component analysis*. New York, USA: Springer.

Kahn, P., Ishiguro, H., Friedman, B., & Kanda, T. (2006, September). What is a human? - Toward psychological benchmarks in the field of human-robot interaction. In *Proc. IEEE intl symposium on robot and human interactive communication (RO-MAN)* (pp. 364–371).

Kaiser, M., & Dillmann, R. (1996, April). Building elementary robot skills from human demonstration. In *Proc. IEEE intl conf. on robotics and automation (ICRA)* (Vol. 3, pp. 2700–2705).

Kambhatla, N. (1996). *Local models and Gaussian mixture models for statistical data processing*. PhD thesis, Oregon Graduate Institute of Science and Technology, Portland, OR.

Kaplan, F. (2004). Who is afraid of the humanoid? investigating cultural differences in the acceptance of robots. *Humanoid robotics*, *1*(3), 465–480.

Kapoor, A., & Picard, R. (2001). A real-time head nod and shake detector. In *Workshop on perceptive user interfaces (PUI)* (pp. 1–5).

Kato, I. (1973). Development of WABOT 1. *Biomechanism*, *2*, 173–214.

Kehtarnavaz, N., & deFigueiredo, J. (1988). A 3-D contour segmentation scheme based on curvature and torsion. *IEEE Trans. on Pattern Analysis and Machine Intelligence*, *10*(5), 707–713.

Kheddar, A. (2001). Teleoperation based on the hidden robot concept. *IEEE Trans. on Systems, Man and Cybernetics, Part A*, *31*(1), 1–13.

Kozima, H., & Yano, H. (2001, September). A robot that learns to communicate with human caregivers. In *Proc. intl workshop on epigenetic robotics*.

Kruijff, G.-J. M., Zender, H., Jensfelt, P., & Christensen, H. (2007). Situated dialogue and spatial organization: What, where... and why? *Advanced Robotic Systems*, *4*(1), 125–138.

Kulic, D., Takano, W., & Nakamura, Y. (2008). Incremental learning, clustering and hierarchy formation of whole body motion patterns using adaptive hidden markov chains. *Intl Journal of Robotics Research*, *27*(7), 761–784.

Lawrence, N. (2005). Probabilistic non-linear principal component analysis with Gaussian process latent variable models. *Machine Learning Research*, *6*, 1783–1816.

Lee, B., Hope, G., & Witts, N. (2006, September). Could next generation androids get emotionally close? Relational closeness from human dyadic interactions. In *Proc. IEEE intl symposium on robot and human interactive communication (RO-MAN)* (pp. 475–479).

Lee, C., & Xu, Y. (1996, April). Online, interactive learning of gestures for human/robot interfaces. In *Proc. IEEE intl conf. on robotics and automation (ICRA)* (Vol. 4, pp. 2982–2987).

Lee, D., & Nakamura, Y. (2006, October). Stochastic model of imitating a new observed motion based on the acquired motion primitives. In

Proc. IEEE/RSJ intl conf. on intelligent robots and systems (IROS) (pp. 4994–5000).

Lee, D., & Nakamura, Y. (2007, April). Mimesis scheme using a monocular vision system on a humanoid robot. In *Proc. IEEE intl conf. on robotics and automation (ICRA)* (pp. 2162–2168).

Lenarcic, J., & Stanisic, M. (2003). A humanoid shoulder complex and the humeral pointing kinematics. *IEEE Trans. on Robotics and Automation*, *19*(3), 499–506.

Levas, A., & Selfridge, M. (1984, March). A user-friendly high-level robot teaching system. In *Proceeding of the IEEE intl conf. on robotics* (pp. 413–416).

Li, P., & Horowitz, R. (1999). Passive velocity field control of mechanical manipulators. *IEEE Trans. on Robotics and Automation*, *15*(4), 751–763.

Lieberman, H. (2001). *Your wish is my command: Programming by example* (H. Lieberman, Ed.). San Francisco, CA, USA: M. Kaufmann Inc.

Liegeois, A. (1977). Automatic supervisory control of the configuration and behavior of multibody mechanisms. *IEEE Trans. on Systems, Man, and Cybernetics*, *7*, 868–871.

Liu, S., & Asada, H. (1993, May). Teaching and learning of deburring robots using neural networks. In *Proc. IEEE intl conf. on robotics and automation (ICRA)* (Vol. 3, pp. 339–345).

Lozano-Perez, T. (1983). Robot programming. *Proc. of the IEEE*, *71*(7), 821–841.

MacDorman, K., Chalodhorn, R., & Asada, M. (2004, August). Periodic nonlinear principal component neural networks for humanoid motion segmentation, generalization, and generation. In *Proc. intl conf. on pattern recognition (ICPR)* (Vol. 4, pp. 537–540).

MacDorman, K., & Cowley, S. (2006, September). Long-term relationships as a benchmark for robot personhood. In *Proc. IEEE intl symposium on robot and human interactive communication (RO-MAN)* (pp. 378–383).

MacQueen, J. (1967). Some methods for classification and analysis of multivariate observations. In *Proc. symposium on mathematical statistics and probability* (pp. 281–297). University of California Press.

Massie, T., & Salisbury, J. (1994, November). The PHANToM haptic interface: A device for probing virtual objects. In *Proc. ASME winter annual meeting, symposium on haptic interfaces for virtual environment and tele-operator systems* (pp. 295–300).

Matarić, M. J. (2002). Sensory-motor primitives as a basis for imitation: Linking perception to action and biology to robotics. In C. Nehaniv & K. Dautenhahn (Eds.), *Imitation in animals and artifacts*. Cambridge, MA, USA: MIT Press.

Maunsell, J., & Van Essen, D. (1983). Functional properties of neurons in middle temporal visual area of the macaque monkey. II. Binocular interactions and sensitivity to binocular disparity. *Neurophysiology*, *49*(5), 1148–1167.

Mayer, H., Nagy, I., Knoll, A., Braun, E., Lange, R., & Bauernschmitt, R. (2007, April). Adaptive control for human-robot skilltransfer: Trajectory planning based on fluid dynamics. In *Proc. IEEE intl conf. on robotics and automation (ICRA)* (p. 1800–1807).

Mayr, O. (1970). *The origins of feedback control.* Cambridge, MA, USA: MIT Press.

McLachlan, G., & Peel, D. (2000). *Finite mixture models.* New York, USA: Wiley-Interscience.

Meltzoff, A., & Moore, M. (1977). Imitation of facial and manual gestures by human neonates. *Science, 198*(4312), 75–78.

Mendel, J. (1995). *Lessons in estimation theory for signal processing, communications, and control.* New Jersey, USA: Prentice-Hall.

Milgram, P., Rastogi, A., & Grodski, J. (1995, July). Telerobotic control using augmented reality. In *Proc. IEEE intl workshop on robot and human communication (RO-MAN)* (pp. 21–29).

Mitchell, T., Caruana, R., Freitag, D., McDermott, J., & Zabowski, D. (1994). Experience with a learning personal assistant. *Communications of the ACM, 37*(7), 80–91.

Moore, A. (1992). Fast, robust adaptive control by learning only forward models. In *Advances in neural information processing systems (NIPS)* (Vol. 4). San Francisco, CA, USA: Morgan Kaufmann.

Muench, S., Kreuziger, J., Kaiser, M., & Dillmann, R. (1994). Robot programming by demonstration (RPD) - Using machine learning and user interaction methods for the development of easy and comfortable robot programming systems. In *Proc. intl symposium on industrial robots (ISIR)* (pp. 685–693).

Muirhead, R. (1982). *Aspects of multivariate statistical theory.* New York, USA: Wiley.

Mussa-Ivaldi, F. (1997). Nonlinear force fields: a distributed system of control primitives for representing and learning movements. In *IEEE intl symposium on computational intelligence in robotics and automation* (pp. 84–90).

Nadel, J., Guerini, C., Peze, A., & Rivet, C. (1999). The evolving nature of imitation as a format for communication. In *Imitation in infancy* (pp. 209–234). Cambridge, UK: Cambridge University Press.

Nakamura, Y., & Hanafusa, H. (1986). Inverse kinematics solutions with singularity robustness for robot manipulator control. *ASME Journal of Dynamic Systems, Measurement, and Control, 108*, 163–171.

Nakanishi, J., Morimoto, J., Endo, G., Cheng, G., Schaal, S., & Kawato, M. (2004). Learning from demonstration and adaptation of biped locomotion. *Robotics and Autonomous Systems, 47*(2-3), 79–91.

Nam, Y., & Wohn, K. (1996). Recognition of space-time hand-gestures using hidden Markov model. In *ACM symposium on virtual reality software and technology* (pp. 51–58).

Nehaniv, C. (2003). Nine billion correspondence problems and some methods for solving them. In *Proc. intl symposium on imitation in animals and artifacts (AISB)* (pp. 93–95).

Nehaniv, C., & Dautenhahn, K. (2000). Of hummingbirds and helicopters: An algebraic framework for interdisciplinary studies of imitation and its applications. In J. Demiris & A. Birk (Eds.), *Interdisciplinary approaches to robot learning* (Vol. 24, pp. 136–161). World Scientific Press.

Nehaniv, C., & Dautenhahn, K. (2001). Like me? - measures of correspondence and imitation. *Cybernetics and Systems: An Intl Journal, 32*(1-2), 11–51.

Nehaniv, C., & Dautenhahn, K. (2002). The correspondence problem. In *Imitation in animals and artifacts* (pp. 41–61). Cambridge, MA, USA: MIT Press.

Nguyen, L., Bualat, M., Edwards, L., Flueckiger, L., Neveu, C., Schwehr, K., et al. (2001). Virtual reality interfaces for visualization and control of remote vehicles. *Autonomous Robots, 11*(1), 59–68.

Nguyen-Tuong, D., & Peters, J. (2008, September). Local gaussian process regression for real-time model-based robot control. In *IEEE/RSJ intl conf. on intelligent robots and systems (IROS)* (p. 380–385).

Nickel, K., & Stiefelhagen, R. (2003). Pointing gesture recognition based on 3D-tracking of face, hands and head orientation. In *Proc. intl conf. on multimodal interfaces (ICMI)* (pp. 140–146).

Nicolescu, M., & Mataric, M. (2003). Natural methods for robot task learning: Instructive demonstrations, generalization and practice. In *Proc. intl joint conf. on autonomous agents and multiagent systems (AAMAS)* (pp. 241–248).

Nicolescu, M., & Mataric, M. (2007). Task learning through imitation and human-robot interaction. In K. Dautenhahn & C. Nehaniv (Eds.), *Imitation and social learning in robots, humans and animals: Behavioural, social and communicative dimensions* (pp. 407–424). Cambridge, UK: Cambridge University Press.

Ogawara, K., Takamatsu, J., Kimura, H., & Ikeuchi, K. (2003). Extraction of essential interactions through multiple observations of human demonstrations. *IEEE Trans. on Industrial Electronics, 50*(4), 667–675.

Ogino, M., Toichi, H., Yoshikawa, Y., & Asada, M. (2006). Interaction rule learning with a human partner based on an imitation faculty with a simple visuo-motor mapping. *Robotics and Autonomous Systems, 54*(5), 414–418.

Otmane, S., Mallem, M., Kheddar, A., & Chavand, F. (2000). Active virtual guides as an apparatus for augmented reality based telemanipulation system on the internet. In *Proc. IEEE simulation symposium* (p. 185–191).

Oztop, E., & Arbib, M. (2002). Schema design and implementation of the grasp-related mirror neuron system. *Biological Cybernetics, 87*(2), 116140.

Oztop, E., Kawato, M., & Arbib, M. (2006). Mirror neurons and imitation: A computationally guided review. *Neural Networks, 19*(3), 254–21.

Oztop, E., Lin, M., Kawato, M., & Cheng, G. (2006, December). Dexterous skills transfer by extending human body schema to a robotic hand. In *Proc. IEEE-RAS intl conf. on humanoid robots (Humanoids)* (pp. 82–87).

Pardowitz, M., Zoellner, R., & Dillmann, R. (2005, December). Incremental learning of task sequences with information-theoretic metrics. In *Proc. european robotics symposium (EUROS)* (pp. 51–63).

Pardowitz, M., Zoellner, R., Knoop, S., & Dillmann, R. (2007). Incremental learning of tasks from user demonstrations, past experiences and vocal comments. *IEEE Trans. on Systems, Man and Cybernetics, Part B*, *37*(2), 322–332.

Park, H., Kim, E., Jang, S., Park, S., Park, M., & Kim, H. (2005). Hmm-based gesture recognition for robot control. *Pattern Recognition and Image Analysis*, 607–614.

Peshkin, M., Colgate, J., & Moore, C. (1996, April). Passive robots and haptic displays based on nonholonomic elements. In *Proc. IEEE intl conf. on robotics and automation (ICRA)* (Vol. 1, p. 551–556).

Peters, J., Vijayakumar, S., & Schaal, S. (2003, October). Reinforcement learning for humanoid robotics. In *Proc. IEEE intl conf. on humanoid robots (Humanoids)*.

Petersen, K. B., & Pedersen, M. S. (2008). *The matrix cookbook*. Technical Report. University of Denmark.

Piaget, J. (1962). *Play, dreams and imitation in childhood*. New York, USA: Norton.

Pook, P., & Ballard, D. (1993, May). Recognizing teleoperated manipulations. In *Proc. of IEEE intl conf. on robotics and automation (ICRA)* (Vol. 2, pp. 578–585).

Popovic, M., & Popovic, D. (2001). Cloning biological synergies improves control of elbow neuroprostheses. *IEEE Engineering in Medicine and Biology Magazine*.

Rabiner, L. (1989, February). A tutorial on hidden Markov models and selected applications in speech recognition. *Proc. IEEE*, *77:2*, 257–285.

Rasmussen, C., & Williams, C. (2006). *Gaussian processes for machine learning*. Cambridge, MA, USA: MIT Press.

Righetti, L., & Ijspeert, A. (2006, May). Programmable central pattern generators: an application to biped locomotion control. In *Proc. IEEE intl conf. on robotics and automation (ICRA)* (pp. 1585–1590).

Riley, M., Ude, A., Atkeson, C., & Cheng, G. (2006, December). Coaching: An approach to efficiently and intuitively create humanoid robot behaviors. In *Proc. IEEE-RAS intl conf. on humanoid robots (Humanoids)* (pp. 567–574).

Rizzolatti, G., Fadiga, L., Fogassi, L., & Gallese, V. (2002). From mirror neurons to imitation: Facts and speculations. In A. Meltzoff & W. Prinz (Eds.), *The imitative mind: Development, evolution and brain bases* (pp. 163–182). Cambridge, UK: Cambridge University Press.

Rizzolatti, G., Fogassi, L., & Gallese, V. (2001). Neurophysiological mechanisms underlying the understanding and imitation of action. *Nature Reviews Neuroscience*, *2*, 661–670.

Rohlfing, K., Fritsch, J., Wrede, B., & Jungmann, T. (2006). How can multimodal cues from child-directed interaction reduce learning complexity in robots? *Advanced Robotics*, *20*(10), 1183–1199.

Rolland, J., Davis, L., & Baillot, Y. (2001). A survey of tracking technology for virtual environments. In *Augmented reality and wearable computers* (chap. 3). Mahwah, NJ.: L. Erlbaum Associates Inc.

Rosheim, M. (2006). *Leonardo's lost robots.* New York, USA: Springer.

Rosheim, M. E. (1994). *Robot evolution: The development of anthrobotics.* New York, NY, USA: John Wiley & Sons, Inc.

Rossi, D. D., Bartalesi, R., Lorussi, F., Tognetti, A., & Zupone, G. (2006). Body gesture and posture classification by smart clothes. In *IEEE/RAS-EMBS intl conf. on biomedical robotics and biomechatronics (BIOROB)* (pp. 1189–1193).

Sakoe, H., & Chiba, S. (1978). Dynamic programming algorithm optimization for spoken word recognition. *IEEE Trans. on Acoustics, Speech and Signal Processing, 26*(1), 43–49.

Salter, T., Dautenhahn, K., & Boekhorst, R. te. (2006). Learning about natural human-robot interaction styles. *Robotics and Autonomous Systems, 54*(2), 127–134.

Sato, Y., Bernardin, K., Kimura, H., & Ikeuchi, K. (2002). Task analysis based on observing hands and objects by vision. In *Proc. IEEE/RSJ intl conf. on intelligent robots and systems (IROS)* (pp. 1208–1213).

Saunders, J., Nehaniv, C., & Dautenhahn, K. (2006, March). Teaching robots by moulding behavior and scaffolding the environment. In *Proc. ACM SIGCHI/SIGART conf. on Human-Robot Interaction (HRI)* (pp. 118–125).

Sauser, E., & Billard, A. (2006a, October). Biologically inspired multimodal integration: Interferences in a human-robot interaction game. In *Proc. IEEE/RSJ intl conf. on intelligent robots and systems (IROS)* (pp. 5619–5624).

Sauser, E., & Billard, A. (2006b). Dynamic updating of distributed neural representations using forward models. *Biological Cybernetics, 95*(6), 567–588.

Sauser, E., & Billard, A. (2006c). Parallel and distributed neural models of the ideomotor principle: An investigation of imitative cortical pathways. *Neural Networks, 19*(3), 285–298.

Scassellati, B. (1999). Imitation and mechanisms of joint attention: A developmental structure for building social skills on a humanoid robot. *Lecture Notes in Computer Science, 1562,* 176–195.

Schaal, S. (1999a). Is imitation learning the route to humanoid robots? *Trends in Cognitive Sciences, 3*(6), 233–242.

Schaal, S. (1999b). Nonparametric regression for learning nonlinear transformations. In H. Ritter & O. Holland (Eds.), *Prerational intelligence in strategies, high-level processes and collective behavior.* Dordrecht, The Netherlands: Kluwer Academic Press.

Schaal, S., & Atkeson, C. (1996). From isolation to cooperation: An alternative view of a system of experts. In *Advances in neural information processing systems (NIPS)* (Vol. 8, p. 605–611). San Francisco, CA, USA: Morgan Kaufmann Publishers Inc.

Schaal, S., & Atkeson, C. (1998). Constructive incremental learning from only local information. *Neural Computation, 10*(8), 2047–2084.

Schoener, G., & Santos, C. (2001). Control of movement time and sequential action through attractor dynamics: a simulation study demonstrating object interception and coordination. In *Proc. intl symposium on intelligent robotic systems (SIRS)*.

Schwarz, G. (1978). Estimating the dimension of a model. *Annals of Statistics*, *6*, 461–464.

Sciavicco, L., Siciliano, B., & Sciavicco, B. (2000). *Modelling and control of robot manipulators*. Secaucus, NJ, USA: Springer-Verlag New York, Inc.

Scott, D. (2004). Multivariate density estimation and visualization. In W. H. J.E. Gentle & Y. Mori (Eds.), *Handbook of computational statistics: Concepts and methods* (pp. 517–538). Berlin, Germany: Springer-Verlag.

Segre, A. (1988). *Machine learning of robot assembly plans*. Boston, MA, USA: Kluwer Academic Publishers.

Segre, A., & DeJong, G. (1985, March). Explanation-based manipulator learning: Acquisition of planning ability through observation. In *IEEE conf. on robotics and automation (ICRA)* (pp. 555–560).

Shimizu, M., Yoon, W.-K., & Kitagaki, K. (2007, April). A practical redundancy resolution for 7 dof redundant manipulators with joint limits. In *Proc. IEEE intl conf. on robotics and automation (ICRA)* (pp. 4510–4516).

Shon, A., Grochow, K., & Rao, R. (2005). Robotic imitation from human motion capture using Gaussian processes. In *Proc. IEEE/RAS intl conf. on humanoid robots (Humanoids)*.

Shon, A., Storz, J., & Rao, R. (2007, April). Towards a real-time bayesian imitation system for a humanoid robot. In *Proc. IEEE intl conf. on robotics and automation (ICRA)* (pp. 2847–2852).

Skubic, M., & Volz, R. (2000). Acquiring robust, force-based assembly skills from human demonstration. In *IEEE trans. on robotics and automation* (Vol. 16, pp. 772–781).

Smith, D., Cypher, A., & Spohrer, J. (1994). Kidsim: Programming agents without a programming language. *Communications of the ACM*, *37*(7), 54–67.

Smola, A., & Schölkopf, B. (2004). A tutorial on support vector regression. *Statistics and Computing*, *14*(3), 199–222.

Song, M., & Wang, H. (2005). Highly efficient incremental estimation of Gaussian mixture models for online data stream clustering. In *Proc. of SPIE: Intelligent computing - theory and applications III* (Vol. 5803, pp. 174–183).

Spexard, T., Hanheide, M., & Sagerer, G. (2007). Human-oriented interaction with an anthropomorphic robot. *IEEE Trans. on Robotics*, *23*(5), 852–862.

Steels, L. (2001). Language games for autonomous robots. *Intelligent Systems*, *16*(5), 16–22.

Steil, J., Röthling, F., Haschke, R., & Ritter, H. (2004). Situated robot learning for multi-modal instruction and imitation of grasping. *Robotics and Autonomous Systems*, *47*(2-3), 129–141.

Sternad, D., & Schaal, S. (1999). Segmentation of endpoint trajectories does not imply segmented control. *Experimental Brain Research, 124/1*, 118–136.

Sturman, D., & Zeltzer, D. (1994, January). A survey of glove-based input. *IEEE Computer Graphics and Applications, 14*(1), 30–39.

Sugiura, K., & Iwahashi, N. (2008, September). Motion recognition and generation by combining reference-point-dependent probabilistic models. In *Proc. IEEE/RSJ intl conf. on intelligent robots and systems (IROS)* (pp. 852–857).

Sun, W. (1999). Spectral analysis of hermite cubic spline collocation systems. *SIAM Journal of Numerical Analysis, 36*(6), 1962–1975.

Sun, X. (2002, May). Pitch determination and voice quality analysis using subharmonic-to-harmonic ratio. In *Proc. IEEE intl conf. on acoustics, speech, and signal processing (ICASSP)* (Vol. 1, pp. 333–336).

Sung, H. (2004). *Gaussian mixture regression and classification*. PhD thesis, Rice University, Houston, Texas.

Sutton, R., & Barto, A. (1998). *Reinforcement learning : an introduction*. Cambridge, MA, USA: MIT Press.

Sutton, R., Singh, S., Precup, D., & Ravindran, B. (1999). Improved switching among temporally abstract actions. In *Advances in neural information processing systems (NIPS)* (Vol. 11, pp. 1066–1072).

Tachi, S., Maeda, T., Hirata, R., & Hoshino, H. (1994). A construction method of virtual haptic space. In *Proc. intl conf. on artificial reality and tele-existence* (pp. 131–138).

Takano, W., Yamane, K., & Nakamura, Y. (2007, April). Capture database through symbolization, recognition and generation of motion patterns. In *Proc. IEEE intl conf. on robotics and automation (ICRA)* (pp. 3092–3097).

Takubo, T., Inoue, K., Arai, T., & Nishii, K. (2006, October). Wholebody teleoperation for humanoid robot by marionette system. In *Proc. IEEE/RSJ intl conf. on intelligent robots and systems (IROS)* (pp. 4459–4465).

Thomaz, A., Berlin, M., & Breazeal, C. (2005, August). An embodied computational model of social referencing. In *Proc. IEEE intl workshop on robot and human interactive communication (Ro-Man)* (pp. 591–598).

Thomaz, A., & Breazeal, C. (2006). Transparency and socially guided machine learning. In *Proc. IEEE intl conf. on development and learning (ICDL)*.

Thrun, S., & Mitchell, T. (1993). Integrating inductive neural network learning and explanation-based learning. In *Proc. intl joint conf. on artificial intelligence (IJCAI)* (p. 930–936).

Thrun, S., & Mitchell, T. (1995). Lifelong robot learning. *Robotics and Autonomous Systems, 15*, 25–46.

Tian, T., & Sclaroff, S. (2005). Handsignals recognition from video using 3D motion capture data. In *Proc. IEEE workshop on motion and video computing* (pp. 189–194).

Ting, J.-A., D'Souza, A., & Schaal, S. (2007, April). Automatic outlier detection: A bayesian approach. In *Proc. IEEE intl conf. on robotics and automation (ICRA)* (pp. 2489–2494).

Tso, S., & Liu, K. (1996, December). Hidden Markov model for intelligent extraction of robot trajectory command from demonstrated trajectories. In *Proc. IEEE intl conf. on industrial technology (ICIT)* (pp. 294–298).

Ude, A. (1993). Trajectory generation from noisy positions of object features for teaching robot paths. *Robotics and Autonomous Systems, 11*(2), 113–127.

Ude, A., Atkeson, C., & Riley, M. (2004). Programming full-body movements for humanoid robots by observation. *Robotics and Autonomous Systems, 47*(2-3), 93–108.

Urtasun, R., Glardon, P., Boulic, R., Thalmann, D., & Fua, P. (2004). Style-based motion synthesis. *Computer Graphics Forum, 23*(4), 1–14.

Vijayakumar, S., D'souza, A., & Schaal, S. (2005). Incremental online learning in high dimensions. *Neural Computation, 17*(12), 2602–2634.

Vijayakumar, S., & Schaal, S. (2000). Locally weighted projection regression: An O(n) algorithm for incremental real time learning in high dimensional spaces. In *Proc. intl conf. on machine learning (ICML)* (pp. 288–293).

Viterbi, A. (1967). Error bounds for convolutional codes and an asymptotically optimum decoding algorithm. *IEEE Trans. on Information Theory, 13*(2), 260–269.

Vlassis, N., & Likas, A. (2002). A greedy EM algorithm for Gaussian mixture learning. *Neural Processing Letters, 15*(1), 77–87.

Vygotsky, L. (1978). *Mind in society - The development of higher psychological processes* (M. Cole, V. John-Steiner, S. Scribner, & E. Souberman, Eds.). Cambridge, MA, USA: Harvard University Press.

Waldherr, S., Romero, R., & Thrun, S. (2000). A gesture based interface for human-robot interaction. *Autonomous Robots, 9*(2), 151–173.

Wampler, C. (1986). Manipulator inverse kinematic solutions based on vector formulations and damped least-squares methods. *IEEE Transactions on Systems, Man and Cybernetics, Part C, 16*(1), 93–101.

Waurzyniak, P. (2006). Masters of manufacturing: Joseph f. engelberger. *Society of Manufacturing Engineers, 137*(1), 65–75.

Whitney, D. (1969). Resolved motion rate control of manipulators and human prostheses. *IEEE Transactions on Man-Machine Systems, 10*(2), 47–52.

Whitney, D. (1972). The mathematics of coordinated control of prosthetic arms and manipulators. *ASME Journal of Dynamic Systems, Measurement, and Control, 94*(4), 303–309.

Wold, H. (1966). Estimation of principal components and related models by iterative least squares. In P. Krishnaiaah (Ed.), *Multivariate analysis* (pp. 391–420). New York, USA: Academic Press.

Wolpert, D., & Kawato, M. (1998). Multiple paired forward and inverse models for motor control. *Neural Networks, 11*(7-8), 1317–1329.

Wood, D., & Wood, H. (1996). Vygotsky, tutoring and learning. *Oxford Review of Education, 22*(1), 5–16.

Wu, G., Helm, F. van der, Veeger, H., Makhsous, M., Van Roy, P., Amglin, C., et al. (2005). ISB recommendation on definitions of joint coordinate systems of various joints for the reporting of human joint motion - part II: shoulder, elbow, wrist and hand. *Biomechanics*(38), 981–992.

Xu, L., Jordan, M., & Hinton, G. (1995). An alternative model for mixtures of experts. In G. Tesauro, D. Touretzky, & T. Leen (Eds.), *Advances in neural information processing systems* (Vol. 7, pp. 633–640). MIT Press.

Xuan, G., Zhang, W., & Chai, P. (2001). EM algorithms of Gaussian mixture model and hidden Markov model. In *Proc. IEEE intl conf. on image processing* (Vol. 1, p. 145–148).

Yamane, K., & Nakamura, Y. (2003). Dynamics filter - Concept and implementation of online motion generator for human figures. *IEEE Trans. on Robotics and Automation*, *19*(3), 421–432.

Yang, J., Xu, Y., & Chen, C. (1993, May). Hidden Markov model approach to skill learning and its application in telerobotics. In *Proc. IEEE intl conf. on robotics and automation (ICRA)* (Vol. 1, pp. 396–402).

Yang, J., Xu, Y., & Chen, C. (1997). Human action learning via hidden Markov model. *IEEE Trans. on Systems, Man, and Cybernetics – Part A: Systems and Humans*, *27*(1), 34–44.

Yeasin, M., & Chaudhuri, S. (2000). Toward automatic robot programming: learning human skill from visual data. *IEEE Trans. on Systems, Man and Cybernetics, Part B*, *30*(1), 180–185.

Yokoi, K., Nakashima, K., Kobayashi, M., Mihune, H., Hasunuma, H., Yanagihara, Y., et al. (2006). A tele-operated humanoid operator. *Robotics Research*, *25*(5-6), 593–602.

Yoon, W.-K., Suehiro, T., Onda, H., & Kitagaki, K. (2006, May). Task skill transfer method using a bilateral teleoperation. In *Proc. IEEE intl conf. on robotics and automation (ICRA)* (pp. 3250–3256).

Yoshikai, T., Otake, N., Mizuuchi, I., Inaba, M., & Inoue, H. (2004, September-October). Development of an imitation behavior in humanoid Kenta with reinforcement learning algorithm based on the attention during imitation. In *Proc. IEEE/RSJ intl conf. on intelligent robots and systems (IROS)* (pp. 1192–1197).

Yoshikawa, Y., Shinozawa, K., Ishiguro, H., Hagita, N., & Miyamoto, T. (2006, August). Responsive robot gaze to interaction partner. In *Proc. of Robotics: Science and Systems (RSS)*.

Young, J., Xin, M., & Sharlin, E. (2007). Robot expressionism through cartooning. In *Proceeding of the ACM/IEEE intl conf. on human-robot interaction (HRI)* (pp. 309–316).

Yu, C., & Ballard, D. (2007). A unified model of early word learning: Integrating statistical and social cues. *Neurocomputing*, *70*(13–15).

Zen, H., Tokuda, K., & Kitamura, T. (2007). Reformulating the hmm as a trajectory model by imposing explicit relationships between static and dynamic feature vector sequences. *Computer Speech and Language*, *21*(1), 153–173.

Zhang, J., & Rössler, B. (2004). Self-valuing learning and generalization with application in visually guided grasping of complex objects. *Robotics and Autonomous Systems*, *47*(2-3), 117–127.

Zhang, Z. (2000). A flexible new technique for camera calibration. *IEEE Trans. on Pattern Analysis and Machine Intelligence*, *22*(11), 1330–1334.

Zoellner, R., Pardowitz, M., Knoop, S., & Dillmann, R. (2005). Towards cognitive robots: Building hierarchical task representations of manipulations from human demonstration. In *Proc. IEEE intl conf. on robotics and automation (ICRA)* (pp. 1535–1540).

Zoellner, R., Rogalla, O., Dillmann, R., & Zoellner, M. (2002, September). Understanding users intention: programming fine manipulation tasks by demonstration. In *IEEE/RSJ intl conf. on intelligent robots and system (IROS)* (Vol. 2, pp. 1114–1119).

Zukow-Goldring, P. (2004). Caregivers and the education of the mirror system. In *Proc. intl conf. on development and learning (ICDL)* (pp. 96–103).

INDEX